Christmas on

Jane Street

Christmas on Jane Street

A TRUE STORY

BY BILLY ROMP

WITH WANDA URBANSKA

Illustrations by Robbin Gourley

William Morrow and Company, Inc. New York

It is the policy of William Morrow and Company, Inc., and its
imprints and affiliates, recognizing the importance of preserving what
has been written, to print the books we publish on acid-free paper,
and we exert our best efforts to that end.

Library of Congress Cataloging-in-Publication Data

Romp, Billy.
    Christmas on Jane Street : a true story / Billy Romp, with Wanda
Urbanska ; illustrated by Robbin Gourley.
        p.   cm.
    ISBN 0-688-16442-0
    1. Christmas—New York (State)—New York.   2. New York
(N.Y.)—Social life and customs.   I. Urbanska, Wanda, 1956–
II. Title.
GT4986.A1R65   1998
394.2663'09747—dc21   98-34905
                                                                          CIP

Printed in the United States of America
First Edition
3   4   5   6   7   8   9   10
BOOK DESIGN BY BERNARD KLEIN
www.williammorrow.com

To our neighbors on Jane Street who have become part of our extended family. Your caring, sharing, and love are the reasons why we have returned each Christmas for more than a decade, forming unexpected bonds and friendships to last a lifetime.—BR

For my mother, Marie Olesen Urbanski Whittaker, for her love of art, life, and Christmas.—WU

We would like to thank Meaghan Dowling, who brought the idea and the two authors together and whose vision became a reality in this book. Thanks also to Charlotte Sheedy and Jody Hotchkiss for their stellar professional guidance and to Laureen Connelly Rowland for her generosity and help. Our spouses, Frank Levering and Patti Romp, deserve enormous appreciation for their invaluable input and endless patience. And special thanks go to our children, Ellie, Henry, and Timmy Romp, and Henry Urbanski Levering, who are, after all, our inspiration.

Christmas on

Jane Street

Prologue

When my daughter lifted the green ribbon on my gift that Christmas morning, my heart started racing. Despite what I do for a living, I've never been big on presents—giving or receiving them. Up until now. This gift meant more to me than any Christmas gift I'd ever given, or received. This gift carried a message I didn't want either of us ever to forget.

Ellie has always been a creative, confident, somewhat headstrong girl. And she's highly intuitive. At least, she's always been able to read me. So when she hesitated between untying the ribbon and lifting the box lid, when her eyes caught mine to be sure I was watching the unveiling, I knew that even before removing the present from the box she had guessed at its significance.

I'm not sure that Ellie—my oldest child and only daughter—understood what she'd given *me* that Christmas. It wasn't anything you could wrap in a box or bundle in tissue and drop into one of these shiny new bags. But I had received something of enormous value. Like most milestones, this one wasn't easy to reach. To be sure, my young daughter had put me through the paces that holiday season. But looking back on it now, every step brought me closer to seeing life in a whole new light. In the end, Ellie helped me rediscover the wonder of Christmas and the sacredness of my family.

It may sound peculiar coming from me—someone as close to being the bearer of Christmas tidings as you'll ever find, short of wearing a red suit and a white beard— that I'd ever struggled with the meaning of it all. But I did. You see, I sell Christmas trees in New York City. The season is short and intense but the rewards are considerable. A good season can make my year.

For the last ten years, my wife, three children, and I have journeyed down from our farmhouse in Vermont to set up a stand on the corner of Jane Street and Eighth Avenue in Greenwich Village. We arrive the day after Thanksgiving and leave on Christmas Eve. Our sidewalk is next to the Jane Street Community Garden, where our freshly cut trees stand like a battalion of toy soldiers, ready to go to work spreading their cheer. For the twenty-eight days that we're here, we try to create a little Vermont Village in Greenwich Village, an oasis of goodwill and greenery amid the city chaos and acres of asphalt.

I'm a mostly modest man, but I will tell you this: I have this gift for matching the right tree with the right customer. The first few years, I didn't recognize it as a talent, then I shrugged it off as nothing out of the ordinary. Finally—and this is where Ellie's gift to me comes in—I began to see my ability as something God-given, something to nurture and cherish.

This is the way I work: I adjust to each person's pace. My customer leads, and I follow. Some people deliberate long and hard over their trees, and I stay right with them. Others point to one and say, "That's it." End of discussion. And that's okay, too.

For me, a Christmas tree is more than a piece of merchandise. Though they all come with the same basic equipment—a trunk and branches and needles—they vary tremendously. Small, scraggly trees need your love and attention, while the tall, imposing ones add grandeur to a foyer or living room. Then there are trees that may not be symmetrical but get you right in the heart because they have soul. They've endured hardship and sing ballads about it, if only you'll listen. The ones I like best have this quality. They're like people with character etched on their faces.

The majority of my customers come looking for help. I put on my listening cap and ask questions. That's where my skills come into play. Do you have high ceilings or low? Is your apartment drafty or warm? Do you have young children or boisterous pets? What ornaments do you plan to hang? When I'm with customers, I'm completely focused on getting them the very best tree possible, the right tree for them.

If they're looking for a woody, aromatic fragrance, I steer them toward the most popular Christmas tree, the Balsam fir. If, however, they want a subtler, sweeter smell, I recommend my personal favorite—the Douglas fir. If they desire a sophisticated tree, tall and regal with strong stiff branches that won't bow under the weight of heavy ornaments, I trumpet the king of trees—the Fraser fir. This royal never sheds. "Leave it up till Easter!" I tease.

I've found that with customers—as with life itself—spirit matters as much as, if not more than, the product. If I can get people talking and laughing, if I can get them into a good mood, they'll buy my tree.

I bill myself as a "full-service stand." It's an accurate description and also opens the door for holiday joking and jesting. I launch into a litany of things I do for free. "I'll deliver your tree," I tell them. "Set it up in your stand, hang your ornaments, and sprinkle on tinsel." If I haven't gotten a laugh by then, I continue. "I'll make your eggnog, wrap your presents, write your Christmas cards." By then, even the most harried New Yorker loosens up. Another thing I've learned is that it's harder to create good cheer in others if you don't feel it yourself.

As anyone knows who's worked even one day selling Christmas trees, you're not just selling the product but the season itself. And that's where I got into trouble on this particular year. Because I am so good at what I do, I have a reputation to uphold. I'm told I have the most successful Christmas tree stand in the city. It's not that I get into a competition with every other stand in the city. I'm not that foolish. I know that as with any true competition you're really only competing with yourself. What happened is that I got so caught up in my goal of doing better than I had the year before that I lost touch with the reason people came to me in the first place. I'd lost touch with why I was trying to bring in all this money.

But I'm getting ahead of myself. This story is about something more basic, more fundamental, than my business and how I separated from, then got back together with, myself. It's about my daughter and how her Christmas dream woke up the dreamer in me, the one I had let fall asleep. I suspect that my story is not uncommon among all of us who celebrate Christmas, other parents who get so caught up in the frenzy of the season that we lose touch with its true meaning. What I learned from my daughter, I now have the privilege of sharing with you. For this, I am eternally grateful.

Now I want to tell you about Ellie and the Christmas on Jane Street that changed my life.

1

Our Village in the Village

 Ellie and I had spent the morning setting up Christmas trees when she sprang it on me. "Why don't we rearrange things for a change? Bring the smaller, scragglier trees from Jane Street, where fewer people see them, to Eighth Avenue and move our Fraser and Douglas firs over there."

At first, I was only half listening. My mind had jumped ahead to the coming month. Watching the noisy stream of cars, trucks, and taxicabs rushing toward us up Eighth Avenue, I could see the entire season unfold in my mind's eye. The pace of my business would start out slowly, build steam during the second week, and peak during the two weekends before Christmas. Tree sales would wind down just before the holiday and be limited to harried, last-minute shoppers and a dwindling number of traditional-ists who set up their trees on Christmas Eve. But Ellie's insistent eyes, fixed on mine, demanded a response.

New York City always seemed to spark new ideas in her—in principle, a good thing. Still, I couldn't help but wish that this particular brainstorm had occurred at another, less pressured time. On the opening day of tree sales, after most of the stand had been set up, I wasn't look-ing for a change in plan. I wanted the stand to be neat, orga-nized, and efficient for business on Saturday. So, while Patti and I try to honor the children's creative impulses whenever possible, I wasn't about to alter the layout of the stand.

"I like your idea, Ellie," I started, trying to be tactful. "But I'm afraid that we're going to stick with things the way they are."

Her brown eyes fixed on mine and for an instant it was hard to read her. Did she think I was becoming too rigid? Could she be right?

Over the years, I've learned that there are certain rules for selling at Christmastime. The first is that people crave predictability. Naturally, customers want to see the same high-quality trees year after year, preferably sold by the same caring hands. But it is equally important for them to know where to find things. Once they learn the lay of the land, they like to go back to the same spot to find their tree—"just like last year."

"But, Daddy," she protested. "You always say you can learn by trying new things."

Though I like to think of myself as flexible and open-minded, the reality is that once I've found a system that works, I like to stick to it.

"True," I allowed. "But people don't like unexpected changes—not at Christmas."

Ellie gave me a look, then, seeming to understand my point, let it go. "Whatever," she said, lifting an unadorned wreath from a stack. Using thin wire, she nimbly fastened on a shiny red bow and some pinecones. Then her mind seemed to leap in another direction. "What time is it?" she asked.

I reached into my pocket, pulled out my aging silver pocketwatch—a family heirloom that I always bring to the city for good luck—and told her the time.

"I was wondering if Emma would be home from school soon," Ellie said.

Emma was Ellie's best friend in Manhattan. Emma lived one block north of the stand on Eighth Avenue in a fancy two-story apartment. The two girls have known each other almost all of their lives. Ellie's first-ever sleepover was at Emma's apartment when the girls were four years old, and their friendship has flourished ever since. During the eleven months of the year when we live in Vermont, Ellie and Emma correspond constantly. They always write in any color other than black or blue, and every letter is sealed with some fancy sticker. Horses, rainbows, and hearts were that season's favorites.

Just as Ellie and Emma observe rituals in their friendship, I have rituals that serve my business. After many years of experience I have come to believe that there's a right way and a wrong way to handle trees. In my view, a Christmas tree is not merely a piece of merchandise, it's something worthy of respect. You start by unloading the truck right. Though the trees are bound into tight versions of themselves for easy travel, once they arrive at the stand, you don't just pick up these sleeping beauties and hurl them into the street. Instead you unload them gently, careful not to damage the branches. I lay mine lovingly on the sidewalk, where they rest until I decide where to display them.

We arrange our trees by size, grade, and species, with the large, premium Douglas, Fraser, and Balsam firs occupying the prime "real estate" along the higher-traffic Eighth Avenue sidewalk and "the Charlie Browns" and other bargain merchandise along Jane Street. (The Charlie Browns are the scraggly Balsam trees that stand three and a half feet or less.) We like to joke with price-conscious customers that the Charlie Browns are trees that only a child—or an imaginative adult—could love. So whenever customers buy one, they feel virtuous, like they have adopted a stray cat from the pound.

Sidewalk space is limited, so I stand most of the trees, still bundled in string, several rows deep and lean them against the tree racks near the street curbs. Then I select a few trees for display to open and bring to life. I lean these against the fence of the Jane Street Community Garden. As pedestrians stroll the sidewalks on Eighth Avenue and Jane Street, they're flanked by trees on their left and right, as though they are walking through an aromatic Vermont forest.

When you're in the Christmas tree business, nothing

feels quite as sacred as opening a tree. If you listen closely, you can almost hear the tree exhale as you snip the strings and loosen its branches. In one fell swoop, that tree is transformed from little more than a lifeless bundle to the grand creature, dancing with branches, that God intended it to be.

This may sound strange, but I try to make a personal connection with each tree that passes through my stand. Each tree has its own individual personality, which I try to coax out when I loosen its branches while holding on to its trunk. There are those special trees you can't help but get close to. I've become so attached to some that I have a hard time letting them go. A regal beauty that deserves pampering shouldn't wind up in an apartment shared by three roommates under the age of thirty. They might go away weekends and be gone on Christmas Day, leaving the tree alone and thirsty. I've seen it happen. Once I even told a customer that the tree she'd selected had been sold when it wasn't true. I just couldn't stand to see her take home a tree that was meant for a loving family. It would have been like witnessing a marriage between two people who were fundamentally incompatible and not speaking up when you had the chance.

A nippy, mid-morning breeze wound its way past the opened trees, ruffling their branches and sending their sweet, woodsy scent out into the street. As Ellie put her finishing touches on the wreaths, I hooked our hose to the spigot across the street and began the water-blasting. With the slight slope to the sidewalks, water ran off and dried quickly. Like the humid forest floor, damp sidewalks make an ideal base for the rows of freshly cut trees. And the humidity seems somehow to blend with and enhance the aroma wafting off the evergreens. Like sweeping up pine needles, hosing down the sidewalks always feels deeply satisfying, even though you know the task will soon need repeating. It is one of my comforting tree-season rituals.

Well-wishers and seasonal neighbors had been stopping throughout the day to welcome us back. Philippe Bonsignour, the owner of Bonsignour, the French gourmet food shop across the street, had been the first to greet us that morning. He was about as close to family as it got in Greenwich Village, which meant pretty close. Every year, Philippe opened the arms of his shop to us. All day long, day in and day out, various members of our family trooped in and out of his place. We went there for water, which we drew from the jugs of bottled spring water stored in his basement. We went there for hot beverages and fresh bread, and sometimes just to warm our hands and feet.

That morning, Philippe delivered two piping-hot croissants swaddled in a red-and-white-checked napkin and a steaming cup of hot chocolate for Ellie and coffee for me. He hugged us both, then gathered up our heavy-duty extension cord and went to work. He carried it across the street, where he plugged it into a power outlet in his basement, then brought it back outside and looped the cord up in the high tree branches over Jane Street and ran it down a lamppost and into our home away from home. That year, like all the others before, I offered to pay him for supplying electricity. That year, like every other, he refused.

"You're good for business," he said, minimizing his generosity. "People come from all over for your trees and, while they're here, they stop in and buy my Christmas treats."

Philippe was right, of course. But the fact remained that his business would have benefited from our being there whether or not he helped us. "Good for business" struck me as a kind of masculine way of masking kindly acts, expressing care without having to reveal deep feelings. I could always spot it because I'd done the same thing myself.

"Are you ready for another great season?" Philippe asked.

"I'm counting on it," I told him, trying to make light of it. What I didn't say—and here's where that masculine "cover" came in again—was that some reversals earlier that year had put me under serious financial pressure. I came to New York the previous day all too aware of the fact that a good tree season could make my year. A bad one would send me deeper in the hole. Before coming down to the city, I had worked myself up into such a state about our finances that, as much as I hate to admit this, I'd become a bit of a grouch. I was short with Patti and the children, and my left eye had started its nervous twitch.

It had gotten to the point that Patti had to sit me down and give me a stern lecture. She said that she and the children wouldn't come to the city if I didn't put my priorities in order. You can rebound from financial difficulties, she argued, but you'll never have a second chance to spend family time together once it slips away from you. The Jane Street operation should serve our family, not the reverse. If she and the kids were down there only to help sell more trees, we would have defeated the purpose. Patti was right, as usual. So before we left, I made a vow not to let myself get so wrapped up in the commercial side of the business that I lost sight of the meaning within each Christmas tree and the broader spirit of the season within our family.

Philippe looked me over hard. "There's something different about you this year. Can you feel it? I can't quite put my finger on it."

The way he was looking at me, I felt as if he already knew all about my money problems. I found his comment unnerving, so I didn't answer him directly. Instead, I responded the way Patti always did when she wanted to be both affirming and noncommittal. "You have good instincts," I said.

He let it go at that. Not pressing the point—allowing me my privacy—was in its own way an act of kindness. The generosity and spirit of the Jane Street community follows me throughout the year. During the eleven months when I'm back in Vermont, I regale my country neighbors with tales of the city. I tell them about the bottomless reservoir of goodwill on Jane Street. Many whose heads are filled with unshakable prejudices can't believe it. They ask me whether I carry a gun in the city and imply that it's not safe for us to take our children there. They have a picture of New Yorkers as cold, impersonal capitalists who don't care one bit about their fellow man.

I have to admit I held some of these stereotypes, too, before my first year selling Christmas trees in Manhattan. That year and every one since, I've come to realize that something unusual happens there at Christmas. The kindness on Jane Street is like a force of nature. It takes on a life of its own in powerful and unexpected ways, pulling along unlikely candidates in its path, and transforming them into happier and more vibrant souls.

It would be hard to catalog the many acts of kindness that we have received over the years, but a few stand out. Each year, Angela from down the street brings us a home-made meal shortly after our arrival. One year, when I had contracted a nasty fever, she took me to her apartment, where her daughter, Donna, nursed me back to health. Another time, Andy, who owns a historic brass foundry, left his storefront office to chase down some thugs who tried to steal a Fraser fir from Patti. Another man,

Andrew, the neighborhood dry cleaner, offers free dry cleaning service throughout our season. Every year a family across the street gives us their fax line for the month so we can run the line into our camper and have a phone. And then there's the local meter maid who's developed a permanent case of forgetfulness about ticketing our camper.

Of all the generosity heaped upon us each year, perhaps the most amazing is the offering of the keys. Every year, a row of hooks over the camper door clangs and dangles with sets of keys to various apartments around the neighborhood. Long-standing friends, like Emma's mother, Anne Abbott, park a set of keys to their apartment with us at the beginning of the season along with the offer to use their place anytime for showers and shaves. But so do near strangers who, when they hear we have no running water in the camper, feel moved to open their homes to us. One year, I counted seven sets of keys on our rack.

Sometimes, as I hear the words coming from my mouth describing the magic on Jane Street to my Vermont neighbors, it's hard even for *me* to believe the truth.

As a Yankee who was raised on the ethic of hard work and self-sufficiency, sometimes I've felt a bit uncomfortable with all the kind deeds coming our way. Like, what did I do to deserve them? And how could I ever repay the many kindnesses? The fact is, you can't put a price tag on generosity or trust.

I gave the matter some thought and came up with a solution. It isn't much, but I make it a point to carry my toolbox with me wherever I go. When I visit the apartments of my Jane Street neighbors for showers and shaves, I always fix leaky showerheads and faucets. If their doors are sticky, I oil the hinges—little things like that. Occasionally, people notice and say something. More often, they say nothing, and that pleases me. I figure I've spotted the problem before they have. I like myself in that role of goodwill elf, stealth handyman.

A lot of people came by the stand that day and greeted us by our nickname, the "tree people." At first, I figured it was because they didn't remember our Christian names. But Victor—the man everyone called the mayor of the West Village—knew our first names and called us the "tree people" anyway, out of affection. "Christmas has officially started!" he declared.

When he came up, I was chatting with Don, the neighbor from across the street who'd just hooked up a phone line for us. I trotted out one of my favorite sayings: "It's another beautiful day in paradise!" Patti and I believe that if you say something often enough, it will come true. If I repeated this often enough, it would certainly calm my nerves about money and let Christmas work its magic on me and my family.

Don and Victor always got into it, kidding around. From inside his canvas bag, Victor produced a bag of walnuts wrapped in cellophane and presented it to me. It was fastened festively with a thin velvet ribbon.

"Walnuts!" I exclaimed. "Ellie's favorite! Thank you so much."

Like most of our customers, Victor was fond of our children. Many had personal favorites, the special child in whom they took an interest and followed from year to year. Some had taken a shine to Henry, the Romp family version of Tom Sawyer—wide-eyed, open-faced, and canny; others always seem to favor the youngest, which that year happened to be Timmy, our towheaded toddler. But Victor was an Ellie loyalist, had been from the start.

"They should be chestnuts," he teased. "But I thought

the 'roasting on an open fire' part might be hard for you to pull off without a fireplace."

I chuckled and called my daughter. "Ellie! Look what Victor's brought!"

Delighted, she could hardly wait to sample one. "Daddy, did we remember to bring a nutcracker?"

"Mom's the chief cook, bottle washer, and kitchen stocker," I said. "Ask her."

Don took one of the walnuts, stuck it in a crumpled paper bag, and laid it on the damp sidewalk. "Who needs a nutcracker?" he said, stomping his heel on the bag. He reached inside and offered the nut meat to Ellie.

"Why didn't I think of that," Victor said, before asking Ellie to pick him a wreath—"the best of the lot."

With an air of authority, she walked down one sidewalk, then up the next, studying the wreaths in their midair lineup. Over the years, I've watched Ellie's role with the public develop. When she was Henry's age, customers would ask her questions just to hear her talk. More recently, they'd begun to take her seriously.

When she returned with an unusually lovely, supple wreath, Victor exclaimed: "You got it! You found the best wreath on the lot!" He paid me, then gingerly pressed into Ellie's hand a tip that I learned later was a five-dollar bill.

Victor just stood there. He closed his eyes and took a deep breath, filling his lungs with the sweet scent of evergreen. We learned long ago that this scent could magically transport customers from Greenwich Village to some other place. For many, it was their memory of home; for others, it was their dream of what home should be.

"Who needs to go to the country for the holidays?" he asked, as if reading my mind. "You bring it here—Christmas on Jane Street."

As he looked from our camper home—parked in its seasonal spot on Jane Street at the corner of Eighth Avenue—to the twin forests of trees, he was unable to suppress a smile.

Don hugged Ellie. "Cut any lumber lately?" he said, teasing her about her jacket.

That day, Ellie and I were "twins" in our matching black-and-red lumberjack shirts and blue jeans. In the morning, Ellie generally waited to dress until after she'd seen what I put on. Dressing alike was one of our most cherished Jane Street rituals. Customers frequently asked, "Is this your daughter?" and seemed pleased to find out she was.

Ellie shrugged, then, spotting Patti, the boys, and Santos the dog in the distance, scampered off to greet them.

"I wonder how long she'll keep *that* up," Don said.

I didn't follow.

"I mean, how long is Ellie going to continue dressing like you? Most kids grow up so fast these days. Especially the girls."

"Oh," I said, dismissing him, "Ellie doesn't have the same interests of other girls her age. She's a tomboy. She's known how to operate an electric drill since she was six. How many six-year-olds would you trust with an electric drill?"

In fact, Ellie often seemed like an old soul, closer in behavior and spirit to Patti and me than she was to her brothers, four and eight years younger respectively. Responsible and hardworking, she had been saving money to buy a horse in Vermont. She'd already settled on a black filly and had picked out a name—Thunder. Her odd jobs back home included helping milk cows and gather eggs for a farmer neighbor at five o'clock every morning. The previous fall, she had grown, harvested, and brought to market several rows of pumpkins.

When Ellie returned, we tackled our favorite job, one of the last in the tree-stand setup. I hoisted her up on my shoulders to nail lights to the top of the tree racks. Whenever possible, I gave Ellie the more responsible job and took the supporting role for myself.

"Look, Dad, no hands," she exclaimed, flinging her arms off the support rack and reaching for the sky. Sometimes her fearlessness brought out my most cautious nature.

I decided to let it go. "All right, Ellie," I said in a let's-get-on-with-business tone. Looking back on that moment, I realize now that she was trying to make a grander statement. It's funny how clearly you can see something in hindsight that you're blind to at the time. All the signs of an imminent change were there at that moment, but my eyes weren't open.

Just as quickly as she'd become the daredevil, Ellie was her serious self again. Expertly, she hammered nails onto the top of the tree racks. Once all the nails were in, she strung the line of lights along and bent the nails back over the line to hold it in place. The string of lights, which ran in a continuous line up Eighth Avenue and down Jane

Street, unified our operation. In the dusk and in the evening, it added the magical appearance of twinkling stars above our small city forest.

With Ellie straddling my shoulders, I felt like the luckiest father alive. In my mind, my daughter was perfect—smart, confident, and strong. She could ride a horse or a bicycle faster than most boys her age, fix a flat on a bike, beat me in chess nine times out of ten, and pass for an adult on the phone. Though I'm not by nature a braggart, Patti continually had to restrain my tendencies to boast about Ellie.

I don't know where it came from, but I do recall having this strong, sudden feeling of gratitude that I could still lift her that year. Since the previous December, she had gained stature, substance, and maturity. I wondered if that would be the last year I'd be able to lift her without straining my back. I wondered if that was the last year I'd be able to call her "my little girl" without seeming like the doting father.

Thinking about how much Ellie had grown made me remember our first Christmas on Jane Street. When we started selling trees there ten years ago, it was on a dare from my wife. That first year we came, Ellie was still in diapers, and Patti persuaded me that it would be fun to spend the Christmas season in New York City. Tree sales were meant to pay our way, no more, no less. It's always fascinated me how life works. You start in one place and end up in another and never could have guessed how significant that starting point would be. That's how it was with our Christmas tree stand. It began on a lark and had become the most important thing our family did every year.

Patti returned from her morning errands, holding Timmy on her hip. Henry was at her side, toting supplies. Patti stood at the corner, looking down one sidewalk and up the other. She surveyed our progress for a few moments approvingly, then disappeared into the camper to begin preparing supper.

Henry helped her carry bags into the camper, then dashed out to where Ellie was hammering. Looking up at his older sister, he tugged on my pants leg. "Let me," he pleaded.

I couldn't trust him with a hammer. If he hit the nail in the wrong place, it could break his thumb. Well aware of Henry's jealousy of Ellie and of her special relationship with me, I racked my brain for some special charge I could give him. "Why don't you get the broom and give the sidewalk its first sweep of the season?" I suggested.

"How come she gets to do all the fun stuff?"

"We've been through this before," I told him. "When you're her age—"

He cut me off. " 'When you're her age,' " he mimicked. "Only thing is, she's *always* going to be four years older than me."

"I have a special job for you," I told Henry. "Something Ellie didn't do till she was older than you are now. But you'll have to wait till after supper before I tell you."

The Corner Tree

Every year for the past ten, we've decorated not one but two Christmas trees. The first is "the corner tree." We never feel fully set up until it's all decked out, standing proud and tall on the corner of Jane and Eighth, directing traffic and distracting busy New Yorkers from their woes. Surrounded by its kinsmen, the corner tree commands a strong and vital presence, greeting people and calling customers our way. Our second tree is, of course, the one we set up for our family Christmas celebration back in Vermont.

I told my son that first night we were in New York what

his special duty would be. "Henry, I want you to pick out the corner tree this year." I delivered the news in an earnest tone, like I was assigning him a most important task.

His eyes flashed with excitement and possibility. He appeared overjoyed but tried hard to conceal it. "You mean, any tree I pick you'll put up. *Any* tree?"

"Yes, Henry," I said. "Any tree." Certain children, I'm convinced, would make the very best attorneys, and Henry was one of that group. Literal-minded, they remember every word you say and will repeat them to you when pleading their case. I added: "As long as you choose the *best* tree for the job. Usually we pick a tall tree, one of the premium Balsams or Frasers."

It was a weighty decision—one he debated all weekend. Henry's choice was never far from his thoughts. He posed a series of questions like, "Is it better if the corner tree is tall or wide?" "Can I choose any kind?" He even drew visitors to the stand, like Philippe and Angela, into the discussion. But I suspected that he was more interested in impressing them with his responsibility than in hearing their opinions. I tried to give him maximum leeway: "Those are your decisions, son, as long as you choose the best tree for the job."

Henry's intensity about the tree was equaled by Ellie's enormous anticipation over her long-awaited reunion with Emma. Ellie had been in town for four whole days and not yet seen her dear friend. She had left numerous phone messages on the Abbott answering machine before getting a call back the previous day. It turned out that the family had been away on a long Thanksgiving weekend, but they would arrive on Monday night—in time for the tree-trimming party!

Ellie had worked alongside me throughout the weekend. She helped me sell and bag trees and even make late-night deliveries. Her initiative was impressive. She made small talk with customers. She told them about her cat, Patches, in Vermont who had kittens twice a year, and she described the horse she was saving up to buy. She had an almost uncanny sense about which customers would want her opinion and around which ones she should remain quiet. She'd also taken it upon herself to help boost wreath sales. Even first-time customers noticed how compatible the two of us were. Several made comments like, "I wish I could get my daughter to help me the way yours does you." When I tallied the daily sales totals, I saw that our teamwork had worked its magic on our sales figures.

In light of Ellie's professional behavior on the stand, it might have been easy to forget that she was still a little girl. Except for two things.

One was Zippy. Zippy was Ellie's stuffed monkey, whom she took everywhere. He was a treasured family member, a favorite of Patti's when she was a girl in the fifties and sixties. He slept on Ellie's bunk at night and spent every meal perched on her lap inside the camper. When she was outside, Zippy nestled comfortably in Ellie's jacket with only his head peeping out to keep an eye on his mistress's doings. The other thing that brought home Ellie's youth to me was the fact that her friend Emma was never far from her thoughts.

As soon as we arrived in town, we'd begun posting fliers made by the kids inviting one and all to our annual tree-trimming party that Monday night. We taped these green, photocopied fliers around the stand and tacked them on bulletin boards in neighborhood buildings. We hung a few on the fence and handed some out to customers and pedestrians. The more the merrier! Come one, come all!

Our family dinner was early the night of the tree-trimming so that we would have plenty of time to set up for our guests. After four days in the city, Patti, the children, and I had settled comfortably into our Jane Street routine. The size of our slide-in camper (seven by eight feet) impressed New Yorkers, who were used to living in small spaces. But what amazed them even more was that Patti *cooked* there. Every day, hot meals sprang forth from her Lilliputian kitchen. And by cooking, I don't mean warming canned soup or reheating frozen pizza, I'm talking cooking from scratch. And we sat down as a family together for at least one, and if we could manage it, two, meals a day.

Equipped with a three-burner gas range, a built-in oven, a shallow sink connected to a ten-gallon water tank, and a maple cutting board for a counter, the kitchen was tight, but just large enough to cook in. An antique wooden tray served as a wall-mounted spice rack that held baking powder, cinnamon, ginger, and the like. Frying pans and dish towels hung from hooks beneath upper storage shelves. Perhaps the greatest novelty in the camper was the icebox. Yes, an *icebox* that kept things cold the old-fashioned way—with ice rather than electricity. But because the icebox stood near our portable electric floor heater, it rarely kept things cold for long. As a result, we never left anything there longer than overnight. Which meant that we lived like the French, shopping every day for fruit and other perishables.

Every morning for breakfast, Patti rolled out biscuits on the cutting board or made bran muffins or potato pancakes. We slathered these with creamery butter and our homemade blueberry or blackberry jam from Vermont. At lunch, she whipped up some heavenly soup or stew, often using leftovers from the previous evening. Dinner was generally lasagna or a casserole, homemade bread, and cooked vegetables. Occasionally, as a special treat, she served fish or venison. For dessert, Patti made pies, cakes, and cookies. Her specialties were apple crisp, gingerbread, and oatmeal-raisin and Christmas cookies.

With pleasant smells wafting from the kitchen, the camper became a hub not only for our family but for customers and neighbors. We made it a practice to share whatever we had with anyone who happened by, to offer a bowl of stew or a fresh muffin to anyone who asked and to many who didn't. Louie, a genial homeless man, was our only "regular." He returned the favor by praising us to the hilt. "I always look forward to Christmastime," he would say, "because the Romps are coming!" Occasionally, mistaking the camper for a food truck, passersby knocked on the camper door and called out: "How much are your hot dogs?" Or: "Do you serve pastrami on rye?"

For this special evening, Patti had prepared a great family favorite—shepherd's pie. We eased ourselves onto the comfortable floral cushions around the dinette table, which stood on a small pedestal. We all held hands and waited for Timmy to start the prayer by bowing his head. I asked the blessing for the food we were about to receive. Being so close didn't drive us to distraction, as people mistakenly assumed, but made for close bonds. Cozy—not cramped—was the word I used to describe life inside the camper.

"You did well," I congratulated Henry, as we dug into the shepherd's pie's mashed-potato crust.

Henry had selected a good, solid Balsam fir tree for the corner. It was a seven and a half footer—slightly shorter than normal—but full-figured and pleasant nonetheless. I knew that once it was decked out in ornaments and finery, it would look even better than it did now.

Henry wanted to know if I would have chosen that particular one. He asked me that afternoon when we were setting the corner tree in its stand. I put him off by asking him what he thought. We had tied the top of the tree to the DON'T WALK/WALK sign so that wind wouldn't topple it. I had added an adapter to the string of overhead lights to accommodate the lights for the corner tree.

"So, Dad," he persisted over dinner. "Would *you* have chosen that tree?"

"I might very well have chosen that tree. You really picked a beauty."

Patti glanced out the window at the tree. "It's well anchored." Then she did a double take: "Look, there's already an ornament on it!"

We all turned to look out the little window above the stove. Light was starting to fade on Eighth Avenue but the corner tree with its single ornament was hard to miss. No family member admitted to having hung an ornament earlier, though it seemed like a possible Henry prank. More than likely, it was contributed by a neighbor who mistook our tree-trimming party for an invitation for guests to bring ornaments.

We ate heartily that night. Actually, something about the entire season made our appetites expand. That was true of everyone in the family, including Timmy, who was now old enough to hold his own spoon and deliver most of its contents to his mouth without spilling. Everyone at the table joked about how much harder Timmy's "workload" was in the city and what it consisted of: learning how to count; learning how to walk a straight line without stumbling; learning how to pet Santos without yanking his fur!

In our marriage, Patti and I are usually compatible. We rarely have trouble with major matters—such as how many children to have, how to raise them, where to live, and how to make a living. And we seldom disagree on matters of principle. It is the little things that cause us to form stubborn positions from which we refuse to budge.

Little things like tinsel. Patti likes tinsel, and I loathe the stuff. My argument is that tinsel is a poor substitute for the icicles it was meant to portray. Patti's argument is that tinsel is festive, fun, and easy to use, so why not?

Rather than have the same argument each year, we came up with a simple solution. We decorate the corner tree on alternate years. Patti takes one year and includes tinsel. I take the next and put on the same ornaments but leave off the tinsel. This particular year belonged to me.

We finished our dinner and moved outside to prepare for the tree-trimming party. We stacked several cardboard boxes together to form a tabletop, which we covered with a red tablecloth decorated with holly leaves. Ellie and Henry helped me carry out boxes of tangerines and large bowls of hot popcorn drizzled with melted butter. Our friend Tom contributed a plate of hors d'oeuvres. Our official landlord, Bill Bowser, who oversees the Jane Street Community Garden, contributed two gallons of apple cider, which Patti spiced and warmed for the carolers.

When a small crowd of guests had arrived, I led a round of songs starting with "O Christmas Tree!" We continued caroling for five minutes before we began stringing the lights on the tree. One of the guests confessed to hanging the reindeer ornament that afternoon when no one was looking. "I shouldn't have told you," she confessed. "Then you might have thought Kris Kringle had left it."

Our gaiety attracted bypassers. One woman, dressed in a fur coat and leopard-print turban, stopped and asked: "Is this a TV commercial?"

I had just about persuaded her to join the fun when Anne Abbott and her children, Emma and Lucie, arrived.

Though shorter and thinner than Ellie, Emma had shot up since the previous year. She looked very pretty in a green wool coat with a black velvet collar and matching hat. For a fleeting moment as I looked Emma over in her coat, I thought of how elegant Ellie would look in such an outfit.

Ellie rushed over to her friend, and they hugged like long-lost sisters. The two boys, Henry and Lucie, picked up where they left off the year before: racing around the corner to scope out the new secret hideouts. The girls smiled and beamed at each other for a long time before settling into a session of serious girl talk. Occasionally, almost as an afterthought, they reached for a piece of popcorn or hung an ornament; but it was clear that the main thing was each other.

Though I stayed busy filling our guests' cups with cider and making small talk, occasionally I overheard snatches of their conversation. I heard Emma tell Ellie about some of the things she wanted for Christmas. I heard her ask Ellie what *she* was getting for Christmas. I cocked my ear to hear Ellie's reply, but a noisy truck lumbering down Eighth Avenue muffled it.

No matter. I *knew* what Ellie wanted for Christmas. I had been laboring for months to make it for her.

It's a Romp family tradition that each child receive one major present each Christmas, a gift meant to last a lifetime. Ellie had been dropping broad hints all year long for what I thought she most wanted. She adored my toolbox and borrowed it repeatedly for one project or another. I had almost completed custom-making her own toolbox, a miniature version of my own. I had begun to fill it with tools, some of which I'd found in duplicate in my shop; others I'd picked up at yard sales. I tried whenever possible to get smaller versions of my tools, ones that would fit her smaller hands as a child and work for her in the future when she became a woman. I planned to fill in with other tools from Garber's Hardware down the street from the camper.

Because I knew it was important for the children to own their things and not have everything drawn into some collective family pot, I planned to make a special mark on each of Ellie's tools. That would be my last job. I hadn't yet decided how I'd mark her tools. Perhaps it would be her initials in green. Or maybe I'd engrave her initials in the tools' handles.

Even if they weren't in my direct range of vision, I was always aware of my children's presence or absence in the space around me. Suddenly I sensed that Ellie had disappeared. Along with the other kids. I found Patti, who, seeing the alarm in my eyes, indicated back toward the camper. When I looked, I could see it rocking with boisterous children inside.

Through the narrow camper window I saw Ellie and Emma and Henry and Lucie hurling pillows at one another. Some good-natured insults could be heard coming from inside as well. It appeared to be a war of the sexes, with Santos uncertain which side to pick. Only Timmy, who had gotten hold of the gold garland and was walking it around the mailbox, remained outside.

Patti was discussing with Anne how the children had grown in the year since they'd seen each other. I chatted with several people and broke loose to make a sale to a man who seemed delighted to have a cup of cider thrown into the deal.

At about that time, Ellie stepped out of the camper purposefully. With Emma right behind her, she made her way up to me.

"Daddy," she said, speaking louder than usual. "I've figured out what I want for Christmas."

I smiled. If this was how she wanted it—to make her

announcement publicly with Emma looking on—I wasn't about to spoil her fun. I looked away for a moment, hoping that I wouldn't give away what I already knew.

"You know what I want?" she asked, then hesitated before continuing. "I want to go to the *Nutcracker* ballet at Lincoln Center with Emma."

I was floored. I looked at her and, for an instant, felt that my daughter was a complete stranger. I said nothing; I didn't know what to say.

She must have taken my silence as a license to continue. "I mean, I don't just want me to go, I want our *entire* family to go, just like Emma and her family."

"Where did this come from?" I asked, my eyes landing on the velvet collar on Emma's fine coat.

"It's my Christmas wish—the one gift I want above all others," she said. I noticed that for the first time since we'd been in New York she was not carrying Zippy.

She would think better of it in the morning, I thought. This was some fanciful idea that would be here today, gone tomorrow. "Why don't you sleep on it?" I suggested, trying to downplay the thing.

She looked at me intensely. "It's not going to change. I want to go to *The Nutcracker*. Can we go?"

Since she was pushing it, I felt compelled to respond. I explained to Ellie that we couldn't afford to shut down the stand for the evening in high season. It would mean a loss of hundreds, if not thousands, of dollars in tree sales. On top of that, you'd have to add in the cost of the evening, which would mean several hundred more dollars in tickets, cab fare, dinner, and the works. It was just not possible.

Ellie seemed not to be listening at all. She looked exasperated, like when a customer she'd spent a long time with just walked off without buying. "But if it's my Christmas wish . . . " she pleaded.

I told her how finances were especially tight that year. I explained the Romp family philosophy. "We only spend money on things of lasting value."

It was then that my daughter gave me a look I'd never seen in her eyes before. For a moment it frightened me, it appeared so hard and distant. All she said was: "Well if you're going to be that way." With that, she turned and, with Emma following behind, returned to the camper.

3

The Candle Stand

It started raining that first full week in December and wouldn't let up. Though the amount of rain varied—it drizzled some of the time and poured at others—the skies remained gray and gloomy. Even when I'm outdoors all day, I try not to let rain or snow dampen my spirits. I always think of my mother's wisdom on the subject. She used to say that she "never met" weather she didn't like. "If the weather makes you unhappy, it's not the weather's fault, so it must be yours."

Actually, rain is good for my trees. They drink it up. Literally. They flourish during a downpour and are pumped up for hours afterward. They seem to stand taller and fuller, and are more alive somehow, the way a person is who's just been well fed or showered with praise.

The rain wasn't the reason, but it was a challenge to remain upbeat that week, given how things had gone with Ellie. *Strained* is the word I'd have to use to describe relations between us ever since the tree-trimming party. She'd grown cold and distant, performing her duties at the stand like a sullen employee rather than my golden girl. She was arriving at the stand later and later each morning. On this particular day, she hadn't come out of the camper at all.

In her defense, the rain had cut into my business, and the few customers I had rarely lingered to pick my brain about tree upkeep or decoration. So I didn't really *need* her. It's just that fourteen hours, standing out in the rain, can get lonesome. In years past, Ellie would always come out to sip hot chocolate and lend moral support.

No more. All through this week, she observed what I was wearing in the morning, then found something as close to the opposite as possible. Pink and purple when I wore brown. Black when I chose white. The message she seemed to want to send was that she was nothing like me. Pointedly trying to get to me, she was succeeding, though I tried not to let on.

Actually, no one except the family saw my clothes those days because they were covered over by the hooded orange rain suit that I wore all day. Amazingly, it kept me dry even during deluges. Designed for commercial fishermen, it was made of heavy vinyl on canvas with snap closures. I wore it along with waterproof boots and a visored hat that kept my eyeglasses dry and clear.

Periodically throughout that slow business day, I stuck my head into the camper, where Patti was cooking and the children playing. I say "stuck my head" because, all decked out in my rain gear, I didn't dare go inside and get the camper wet. For meals on wet days, I either had to extricate myself from my rain suit, which was a bother, or Patti would hand me a hot plate covered over in aluminum foil. If it was an off hour and a table was free at Bonsignour, across the street, I'd take my meal over there. As I ate, I kept an eye on the stand out the giant plate-glass window.

Ellie had been holed up in the camper all that day making Christmas cards. Every year, Patti's mother sent the children a care package containing colored paper and envelopes, glitter glue, and Magic Markers from which they spun out magical Christmas creations. Sometimes they'd attach a few Balsam needles to the cards for aroma and effect. My mother-in-law addressed the package to "The Romps, Christmas Tree Sellers, Corner of Jane Street and Eighth Avenue, New York, New York, 10014," and it never failed to reach us, which gave everyone a great kick. With the rain pounding the tin roof, the camper felt cozy and cavelike. Like white noise, the rain somehow muffled the sound of the street.

When I stuck my head in and tried to make conversation, Ellie rebuffed me with one-word answers. "How are the cards coming?" I'd ask, and she'd say, "Fine." Or I'd try, "Getting hungry?" and she'd answer, "Nope." The subject of the *Nutcracker* ballet hadn't come up again between Ellie and me since the tree-trimming party. Had I not known her better, I would have assumed that she'd forgotten all about it. But her behavior was so unlike the girl I knew that it was obvious something was wrong.

The boys had mixed Henry's toy soldiers—a present from Jane Street friends the previous Christmas—with my tools. Soldiers were setting up camp inside an area on the floor bounded by a wrench and hammer. Command headquarters appeared to be the inside of my toolbox. My tools weren't toys, and I didn't appreciate their being used that way. But, as a parent, you have to choose your battles. And I didn't feel like getting into it with Henry just then. Ellie was as much as I could handle. Which may have been why Henry had chosen that moment to dive into my toolbox. He knew that he could get away with it that day.

I tuned in to the weather station on my transistor radio. It was then that I heard a weather advisory: the temperature was going to plummet into the mid-twenties that night. Though it was still early afternoon, you could feel the cold knocking at your bones, frigid fingers poking your body. My thoughts immediately turned to the trees—especially the Douglas firs.

The most delicate of the trees I sell, the Douglas firs have a finer needle than the Frasers and Balsams and are less hardy after being cut. If they're wet when it freezes, they lose color, shed needles rapidly, and appear limp and lifeless. In short, they become unsalable. Thousands of dollars of inventory was standing in the balance on that tree rack. "Ellie," I said in a businesslike way. "I need your help."

She waited for just a moment before acknowledging me. When she did look up from the page, she squinted at me, as if she didn't want to see my entire face but only a fraction.

"I need your help," I said. "Covering the Douglas firs."

"Now?" she asked in a tone that suggested that "now" hardly seemed possible.

How could I make myself any clearer? "N-O-W," I spelled out. My tone was terse and definitive.

"Dad," she groaned. "I don't want to get wet."

I didn't have a lot of sympathy. "Know what?" I said, trying to make light of it. "Your clothes will dry, and your skin's waterproof."

Usually she laughed when I trotted out familiar sayings, but this time, she responded with a resigned smirk. She pulled on her rain slicker and stepped out of the camper behind me. In the cab of the truck, she held Santos back while I leaned the bench seat forward to remove the tarp from behind.

At the Douglas fir rack, Ellie held the tarp off the ground as I tied one end to the base of the rack. The plan was to hoist her up on my shoulders so that she could help ease the tarp over the top of the trees. (Then, when there was a break in the traffic on Eighth Avenue, I'd scamper over to the street and, with lightning speed, tie the other end down.) Before lifting her up, I reminded her to tuck in any loose branches to prevent breaking the limbs. "I know," she answered dismissively.

We were back on the sidewalk when she brought it up again. "Remember what I told you the other night, what my Christmas wish was." She waited for me to nod my head yes before going on. "I was wondering if you'd thought about it—taking the whole family to *The Nutcracker* this year."

"Of course, I remember," I said, watching rain stream down on her hatless head, soaking her long brown braids. I didn't admit it but the fact is I'd questioned more than once since then if I had been too harsh in my judgment about *The Nutcracker*. But I remained convinced that in principle I was right. We really couldn't afford it, and I didn't want to set a precedent here. "It's just not possible for us to do things like that in our financial situation, and certainly not at this time of year," I explained. "We're just like those apple growers in Vermont during the fall harvest season. They can't just leave the stand on a whim."

"But when you ask for a Christmas present, aren't you supposed to ask for what you *really* want?" she said.

"And isn't that what Christmas is sort of about—giving and getting what you really want?"

I thought about it for a moment, then responded as best I knew how. "You can *ask* for what you want, but if it's out of the giver's reach, you shouldn't be sullen about not getting it. And, no, I don't think 'getting what you really want' is what Christmas is about. It's not a material holiday."

"Then why are you trying so hard to make all this money—if it's not a material holiday?"

"There's a difference between keeping food on the table, a roof over your heads, and clothes on your backs, and spending all your money on one Christmas gift," I said. "Someday, when you're a parent, you'll know what I'm talking about."

Something I said must have triggered a reaction in her because her lips started puckering like she was starting to cry. In the rain, though, it was hard to know for sure.

"In most families, Christmas is for children," she said, her voice quivering. "But we always do things the way you want to do them. We're never even at home for Christmas. I mean, we are home but we get there so late, we're never home on Christmas Eve. By Christmas Day, everyone's so tired from the drive home that nobody has any fun."

That one stung, but I knew for a fact it wasn't true. Ellie loved our Christmas celebration in Vermont; either that or she was a great actress. "You don't mean that, Ellie," I said. "You don't mean to say that you never enjoy Christmas." It *was* true we'd always tried to keep the stand on Jane Street open as late on Christmas Eve as possible, to pick up any last-minute tree and wreath sales. I never wanted to leave much unsold inventory behind.

"I do mean it," she said. "And I'm not going to enjoy it this year unless *I* get what I want for a change." She was starting to sound like a stranger, not the daughter I thought I knew.

But how could I make my situation any plainer to her? This was our business; this was how we paid our bills. And her timing was terrible. After a week of sluggish sales, our situation wasn't looking particularly rosy. "I don't think you understand what I'm up against, Ellie," I pleaded.

"Dad, Emma gets to go to *The Nutcracker* every year. I just want to go once." It certainly would have been easier just to peel off five twenty-dollar bills from my wad, hand them over to her, and be done with it. And I have to admit that at some level, I didn't even understand my own resistance. But there was something about her desire to go to the ballet—with all its glamour and glitter—that I found deeply troubling. In my gut, I felt that if I let her go, this might be the beginning of a whole new way of life for her. I feared that my precious daughter would turn her back on our family's simple life in favor of something more worldly.

"Ellie," I started.

She cut right in. "I've made up my mind, Dad. I'm going to go to *The Nutcracker* with Emma—whether or not the rest of you go with me." She glanced at my leg, where, under the rain suit, I kept my roll of money. "I'd rather go alone anyway." Then she played a card I hadn't considered. "I want to get some money from my savings account to buy a ticket."

Ellie had a bank account back in Vermont that she'd opened to save to buy a horse. "I can't let you do that," I said. "That money is for your horse."

She gave me an uncomprehending look.

"I would feel irresponsible as your father to let you spend so much money for a night out on the town. The evening might set you back a hundred dollars or more. It will make you happy for one night and then it will be over. A horse would be your friend for years."

She looked at me fiercely. "But it's *my* money," she said. "I can't believe this. You're not going to let me spend my own money?" The way she looked at me was as if I were the most unreasonable man who'd ever lived.

"It's your money," I said. "But I'm your father. I'm supposed to prevent you from making foolish decisions. That's what parents are for. When you're eighteen, you can make up your own mind about money." I considered repeating how tight money was for us this year, but I held back. I realized that Ellie had mentally separated her finances from ours. At a certain point, she clamped her mouth shut. And I wondered if I'd finally persuaded her to my thinking.

"It's not fair," she said in a low voice that sounded drained, even slightly defeated. Then she stepped back inside the camper, slowly drawing the door back behind her.

The next morning, our second Friday in New York, was one of those glorious days when, after a long period of rain and drizzle, everything sparkled like newly washed windows. The sun was shining and all the world—the buildings, sidewalks, storefronts, and even the pedestrians—seemed to gleam. Maybe it was my imagination, but I could have sworn that God himself had turned down the volume control on the traffic noise in the street. There were fewer honks and rapid, screeching halts and less shouting on the street. Even our trees seemed perky, like schoolchildren after a vacation who are raring to get back to class. The best news of all was that, thanks to the protection of the tarp, our Douglas firs had stayed dry enough to make it through the cold night without freezing.

Ellie climbed down from her bunk bed early that morning. It delighted me to see that her old spark had returned. She volunteered to walk the dog and slipped out of the camper.

While we were in New York City, the cab of the truck was Santos's domain, his doghouse away from home. Most of the time, he slept not in the camper with us but in the cab, where he had more room to stretch and scratch. When he saw Ellie approaching, he reared up on the bench seat and began panting in anticipation. He adored Ellie, who in her mood had ignored him all week. Santos stood obediently as she fastened the leash on his collar. Though he got between two and four walks a day, the morning walk was his favorite by far.

Santos was just a year old, almost full grown. We'd brought him home the previous Christmas Eve from an older Irish couple who worked as building superintendents on the Upper West Side and bred shepherds in their courtyard. As luck would have it, we had gotten the pick of the litter. Santos had been selected first by someone else and then returned when his would-be owner discovered that his lease forbade pets. We named him after a loyal and dependable friend in the hope that he too would come to possess those traits. It didn't occur to us until later that the name also fit with his Christmastime arrival into our household. A lot of customers mistakenly called him "Santa."

Over a breakfast of potato pancakes, biscuits, and blackberry jam, I commented that with this beautiful weather I was expecting "big business." My hope was that Ellie, having cast off her gloom, would return to my side to help wait on customers, bag trees, and make change.

She answered that she was busy herself.

"What are you busy with?" I wanted to know. Ellie smiled coyly and said not a word.

I looked over at Patti, whose playful expression told me she was in the know. "Just you wait," my wife said.

Ellie stepped out of the camper and made a beeline for the boxes of sawed-off tree stumps stored beneath the camper. When I went outside, it was obvious she was sorting through them.

"I'm going to sell Christmas candles," Ellie said. "From now on, would you mind cutting the stumps as evenly as possible?" She was planning to use the stumps that I cut off from the bottom of every tree as bases for Christmas candles. She set off for Woolworth's on Fourteenth Street (I learned that Patti had loaned her some money) and returned later with several sacks full of supplies, including small utility candles, red ribbon, paint, and glue.

How could I possibly object? She was not only showing enterprise but recycling tree stumps that were normally thrown away. Ellie brought out a folding chair from the truck cab and set up her operation outside the camper. Using my battery-operated electric drill, she drilled holes into these stumps to hold the candles. She then dressed the stumps with bows and fir sprigs and laid them out on a little metal folding table. She painted a sign in green and gold lettering: "CHRISTMAS CANDLES: $3 EACH/TWO FOR $5."

Within an hour of hanging out her shingle and spreading our her wares, Ellie had already collected eight dollars, which she tucked into her bluejean pocket. The more I thought about it, the happier I grew over this sudden development. Clearly, Ellie's mind had turned from this foolishness about *The Nutcracker* to something worthwhile. A business of her own, something she could build on. Still, it bothered me that she was so absorbed in her candles that she didn't have much time to help me.

When I had more customers than I could handle, I called for her help.

"In a minute, Dad," she responded, continuing to chat with one of her own customers.

I watched out of the corner of my eye as she completed that sale. When several more tree customers arrived and began milling about the "action zone"—the point where the two sidewalks converged into a giant triangle and most of our sales took place—once again, I called for her help. If someone didn't attend to them soon, I was afraid we'd lose their business. One man, dressed in an expensive Italian suit, kept checking his watch.

Just as her customer walked off, another couple appeared and began admiring her candles. Ellie greeted them as if she had all the time in the world. "It's another beautiful day in paradise," she said, borrowing my well-worn line. It was flattering to be imitated in this way. But she wasn't helping me with the tree sales, something we'd always done together. I considered ordering Ellie to come and help out. After all, that impatient customer was probably good for a fifty- or one-hundred-dollar sale, whereas Ellie would spend the same amount of time for a three-to-five-dollar sale. But remembering how unpleasant the previous days had been when she was so mad at me, I held my tongue. I turned from my own customer, who was deliberating between a six-foot Balsam and a seven-foot Fraser, and greeted the impatient customer. "I'll be with you shortly," I said. He glanced at his watch once again and told me that he'd be back later. I knew that there was at least a 50 percent chance that we'd lost the sale.

I was holding a wreath when I heard Ellie tell her customers why she'd gotten into the candle business: "I'm saving money to buy a ticket to *The Nutcracker.*" I was so stunned that I nearly dropped the wreath. The intensity of her desire to see this ballet was beyond my comprehension. Where was it coming from? Was she more attracted to the bright lights of the city than to the quiet country life that we led?

I didn't know what else to do, so I reflexively fished into my pants pocket and pulled out my silver pocket watch. It was a Hamilton railroad watch that had belonged to my grandfather, Henry Romp. My grandmother had given it to me the year he died, when I was eleven. As the only boy in a family with four sisters, I alone could carry on the Romp name, she had told me. Amazingly, more than three decades later, I still had it, used it, and kept it in good repair. It was a family treasure that I planned to hand down to Henry, when he came of age. Whenever I looked at it, it linked me to the generations who'd come before me and the ones that would follow after. It made me see that whatever was bothering me at the moment would pass. Suddenly *The Nutcracker* seemed less of an issue than it had before. After all, Ellie was working for what she wanted—a philosophy I'd been preaching for years. Then I had a flash.

Someday, Ellie herself might be running my Christmas tree stand.

I walked over to her stand and laid my hands on her shoulders. "Ellie, you're doing a great job!" I said. "Now I know that if I get into a bind, with your skills, you can run my business!"

But her shoulders felt rigid to my touch. Instead she pulled her money from her pocket and began counting it.

"Looks like you're doing well for yourself," I said, thinking perhaps she'd tell me how much she'd taken in. But she didn't volunteer her totals. She didn't say much of anything.

Once that day's rush had died down, she put away her candles and went directly to Emma's apartment for dinner.

Night on the Town

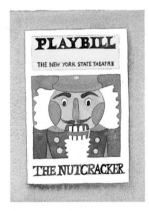

My Christmas trees had been marching off Jane Street into their new homes in record numbers these past two and a half weeks. And Ellie had already made and sold three batches of two dozen candles each. Not one to be outdone by his big sister, Henry had started his own business. At first, his scheme was to sell candles identical to hers and undercut her price. When I put my foot down, we came up with the idea of his selling boughs.

Henry sold the branches that we snipped off trees when we pruned cumbersome boughs or trimmed the trunks. He gathered these up into a giant bundle and invited customers to pick out as many as they pleased. Then, according to the size of the bundle, he'd set a price that usually ranged between fifty cents and a dollar. He tallied his seasonal total every day and let us all know what it was. Since starting, he'd already made sixty-three dollars. And, unlike his sister, he had no expenses. For her part, Ellie refused to divulge her earnings, but I could tell she was doing well.

Probably because our totals at the Christmas tree stand kept rising and were starting to amount to something substantial, my spirits had been climbing with them. I had to remind myself not to get giddy about how well we were doing this season. Even with one good sales month, our family finances were far from being out of the woods. And I had to keep pushing myself to maximize for sales the rest of the season.

Thankfully, we had moved out of the rainy spell of early December into cool, crisp weather that was ideal for selling trees. Any retailer will tell you that you need a nip in the air to put Christmas in people's minds. My worst season ever was the December when the weather was unseasonably warm. Christmas just crept up on everyone without warning.

Whenever possible, I make it a point to do little favors for people, to spread goodwill from our corner. I do these things not just for my customers but for regulars on the street, too. One of my favorites is hailing cabs for the people who travel uptown every morning. I always get a kick out of pointing to where I want the cab to stop, opening the door, and settling my charge into the seat, like a concierge at some fancy hotel. Everyone—especially those who're unaccustomed to this kind of service—is delighted.

Once I did this for a man I'd seen on the street only a few times but we'd never talked. I hailed a cab for him and opened the door precisely as he stepped up to the curb. Not one second of his time was wasted. He seemed surprised, then tried to reach into his billfold to tip me. I waved away his money. "I'm doing this for our public relations department," I said. "Just tell your friends where to buy their tree."

He looked puzzled for an instant. Then he smiled as it

dawned on him that I *was* the public relations department.

My cheerful mood at the stand and in the camper helped raise my family's spirits. Though relations with Ellie had not returned to the idyllic state of the years before, they had improved. I still didn't approve of what she was saving for, but couldn't argue with her enterprise. She'd stopped giving me the cold shoulder and even asked my opinion a time or two about such things as whether to buy fancy bags for her customers or to reuse the old ones that we saved.

Leaning against the camper door, making small talk with Patti, I was sipping my umpteenth cup of coffee that day. Philippe had kindly offered me all the coffee I could drink on the house for the season. "If you don't drink it, I'll throw it away," he said, characteristically downplaying his generosity. I did my small part by bringing my own "container," a thermos mug with handle and drink spout. Bonsignour's coffee was strong and addictive and kept my blood pumping. But I knew I was drinking too much as it was putting me on edge.

Ellie returned to the camper from her regular visit to

Emma's, gleaming like a newly decorated Christmas tree. It was clear from her manner that she was carrying a big piece of news or had just achieved some major triumph. "Guess what!" she repeated several times. "Guess what the Abbotts bought me for Christmas!"

"Ellie," I started. She could tell from my tone what I was getting at—that as a family we'd decided not to open gifts until we were all together on Christmas Day. It was a decision that we had made several years before so that we could savor each gift. Lots of our Jane Street friends wanted us to open their gifts when they delivered them so they could watch. But when we explained our family ritual, people understood. What was it about this season that was affecting my daughter this way? It seemed like each time I turned around, Ellie was testing the limits.

"But, Dad," she started, "I didn't *open* this one. I mean the Abbotts bought me a ticket for *The Nutcracker* as a *Christmas present!*" The way she said it was as if this were the most wonderful Christmas present she had ever received.

Next thing I knew, Patti—still wearing her oven mitts—was hugging Ellie and the two were dancing around the camper as if she'd just achieved something of significance, like winning a ribbon at the horse show or coming in first in a bike race. "Ellie, this is wonderful!" Patti exclaimed. "Your dream come true for Christmas!"

In their excitement, neither my wife nor my daughter was paying me the least bit of mind. I knew I couldn't blame them for leaving me out of the loop. I'd actually removed myself by taking such a strong and vocal stance against *The Nutcracker*. It's hard to celebrate a victory when you were on the other side of the battle to begin with.

The best thing to do, I reminded myself, was not to make too big a deal out of this, to go on with the business at hand. Ellie and Patti stepped outside the camper, where I'd returned to sweeping.

"When are you going, Ellie?" Patti asked.

"Tonight! I've got to get *ready*." The word *ready* fell out of her mouth with a loud thud, like something urgent and immediate.

They started talking about what Ellie was going to wear. The discussion was short. It turned out there was just one candidate—her denim jumper. It was her only garment with a skirt; the only other things she'd brought with her from Vermont were pants and jeans. She'd wear the jumper with white panty hose, and her work boots.

"Is it clean?" Patti asked, concerned.

"I guess," Ellie started, then her mind moved elsewhere. "Mom, I don't have the right outfit."

"We don't have time to go shopping for something else," Patti said. "Not if you're going tonight."

"I've got money."

"You know what, Ellie," Patti said, taking Ellie's long braid in her hand and rubbing it the way you would a cat to make her purr. "You're going to look wonderful. Because it's you—not your clothing—that's going to *The Nutcracker*."

I'm glad no one asked my opinion. Because buying a new outfit for one evening was the kind of extravagance that really rankled me and cut to the heart of my objection to this whole enterprise.

"So what do *you* think?" Ellie asked, facing me directly. I couldn't tell if she was looking for my approval or wanting me to acknowledge her sudden good fortune.

"About what?" I asked.

She looked at me. "You know what. *The Nutcracker*."

"Haven't we been over this?"

"I mean Anne Abbott giving me a ticket as a Christmas present."

I think she wanted me to congratulate her for achieving her goal so effortlessly. The problem was that she was getting what she wanted without having to work for it. Something about this sudden turn of events bothered me. "To tell you the truth," I started. "I'm not crazy about this, and I'll tell you why. You make a commitment to start a business and sell candles to save up for something. And that's fine. I can admire that—even if I don't approve of your end goal. But then along comes some fairy godmother who falls out of the sky and waves a magic wand and gives you this thing you want. I just don't want you to think you can get the things you wish for without having to earn them first; life just doesn't work that way."

She looked at me wryly, summoning all the wisdom of someone three times her age to her face. "Sometimes it does."

"At this point, you need to learn important life lessons," I told her. I knew what I was saying sounded like a lecture, but it was something she needed to hear. "You need to learn the value of hard work, of paying for what you want. If you work for two weeks and spend all your earnings on one night on the town, you may think twice about it next time around. But if something just drops in your lap, you don't recognize its value. That is what I object to."

It was hard to read Ellie just then, to see how she was reacting to my speech. It was clear that she was listening intently, taking in every word. A customer examining the Balsam firs at the far end of the stand on Eighth Avenue caught my eye; as I turned to look, Ellie's eyes followed. Breaking the cardinal rule of retailing, I decided not to greet the customer immediately but to stay with Ellie until I got through to her. "The point is it's not a good thing for you to expect the things you want to drop into your lap. I want you to grow up to be self-sufficient, not dependent on others. You need to learn that only you can make your dreams come true. If you rely on others to do that for you, I'm afraid you'll end up with disappointments."

"This ticket is costing you nothing," she answered. "I don't know why you're being so mean."

"Mean—I'm not being mean. I'm trying to help you." What I didn't tell her was that I was also worried sick that my little girl was developing a taste for fancy things. Things I couldn't provide. A world I didn't travel in. Though I'd never personally developed such tastes, I knew that life's luxuries could be a terrible trap.

Ellie bent down and picked up a pinecone that had fallen loose from a wreath and laid it out of harm's way on her candle stand. "You don't seem happy for me," she said. "Not one bit." With that, Ellie turned and walked off to Emma's apartment, her small canvas bag holding her modest clothing slung over her shoulder behind her.

As soon as she was out of sight, Patti was right in front of me. Timmy was holding her hand. She had one of those "I've-got-a-score-to-settle" looks on her face. "I have one question for you, Billy. Why are you making a federal case out of this? Why can't you let Ellie have her fun? Now you've sent her off with a dark cloud hanging over her— the cloud of your disapproval."

"That's not one question—it's two," I said, trying to make light of it. But she didn't laugh. She wasn't about to be distracted.

"Billy," she said, exasperated.

"You want to know why I'm having trouble with this? This whole *Nutcracker* thing came out of nowhere. Since when was Ellie interested in the ballet? She's a *tomboy*. She likes horses and bikes and trees—not the ballet and fancy clothes and elegant evenings on the town." Patti was looking at me incredulously, and I realized that my voice had gotten loud. I was practically shouting.

Patti's look told me she was about at the end of her rope. "Someone's behavior only bothers you if there's something in yourself you don't like."

"Patti," I said. "What are you talking about? I have no interest in the ballet."

"What is it about this situation that's getting to you?" she asked. "Maybe *you're* the one who wants riches; maybe that's why you're so obsessed with money this year."

I rolled my eyes. "You know why I'm worried about money. I'm trying to dig us out of a hole."

"Maybe we should have bought her something fancy to wear," she said. "It's her big night, and she's not dressed right."

"Patti," I said. "You're starting to sound like Ellie now. Next thing I know, you're going to want to go to *The Nutcracker*."

She looked at me fiercely. "My only concern is Ellie and giving her a good childhood. Happy memories to look back on." There was a touch of bitterness in her voice, as if to say that the two of us were working at cross purposes. Seeing that our discussion was going nowhere, she scooped up Timmy, crossed Jane Street, and went into Bonsignour.

Out there alone on the pavement with just my trees, I started mulling over the events of the day. I was beginning to wonder if Patti might be right about Ellie's venture to Lincoln Center and even her outfit. I wondered if I'd spoiled the evening for her. And I couldn't shake what Patti had said about me. That maybe *I* was the one who wanted riches and was protesting too much about Ellie. Could it be that I, too, was drawn to the trappings of city life? That I was enjoying the fine coffee and bread and foods from Bonsignour and gifts from our customers a bit too much? Was there something other than the good money to be made that kept luring me back to Manhattan year after year?

More than once that season, I wished I'd never said a thing about *The Nutcracker*. Not that my being disappointed and hurt didn't make perfect sense. I'd spent half the year making my daughter a gift that I thought she wanted most only to find out that she wanted something else more. But I also saw that I'd painted myself into a

corner on this *Nutcracker* situation—a corner that would
be hard to step out of. Then I remembered a resolution
I'd made to myself when Ellie was born—that my relationship with her would be without judgment or conflict,
flawless from beginning to end. And it made me sad to
think that we'd already had our first major row.

A striking, expensively dressed customer interrupted
my thoughts. She'd been leaning into the Balsam fir
stand, inhaling the sweet scent of the trees.

"Happy winter solstice!" she called out. Most customers wished me a "Merry Christmas" or its politically
correct cousin, "Happy Holidays." But come to think of
it, she was right; it *was* December 21—the winter solstice.
It was a sign of how wrapped up in the commerce of the
city I had become that it was after four o'clock and it had
never once occurred to me that that day was the shortest
one of the year.

"Are you having a happy one?" I asked.

"Hanging in there," came her response. It never failed
to surprise me how many people who're considered suc-
cesses in this life—well-dressed professionals with lots of
money—who, if you ask them how they are, give you an
answer that falls short of the ideal. They'll say they're

"surviving," "getting by," or "hanging in there." Few give you the impression that they're like my trees at the moment of opening—bursting with energy and life.

"Can I help you find a tree?" I asked, pegging her as a candidate for one of our first-class Fraser firs.

"Just looking," she said.

She didn't buy a tree but instead selected a plain Balsam wreath. She then cast her eyes to the small stand adjacent to the camper where Henry was sitting on a box crate, peddling his wares along with Ellie's, for which, in her absence, he charged a commission. After carefully considering several of the candles, she selected one with a thick base.

"Did you make these?" she asked Henry.

"No, my sister did. I'm selling them for her. She's going to see *The Nutcracker*."

"Well aren't you the diligent worker!" she exclaimed. "What's your name?"

When he told her, her eyes brightened and a smile started to work on her mouth. "My grandfather was named Henry. You don't hear it much in boys your age."

"My great-grandfather was named Henry, too," he said. "That's who I was named for and someday when I grow up, I'm going to get his watch."

She scooped up a handful of branches from Henry and paid him for them and the candle.

"Come back and see us!" he said. The way he said it I could tell he really meant it.

Sales had been brisk all that evening. I helped match several customers with their trees, including a young father who had come to select a tree with a little redheaded girl in tow. The girl, who told me her name was Erica, was a bit younger than Ellie. I soon learned that she was his stepdaughter. He told me: "This is our first Christmas together. Her mother and I were married on June twenty-sixth." I handed Erica a Balsam bough for good luck.

Just then, another little girl standing on the corner, right across the street, caught my eye. She looked like a beautiful little princess, all decked out in party clothes. She had long golden hair like Ellie's. It took a moment— maybe more than a moment—for me to realize that it *was* Ellie.

She was wearing a black velvet jumper, a frilly white blouse with puffy sleeves, and had a black cape flung over her shoulders—an outfit that seemed tailor-made for her. Every accessory was right. She had on white stockings and black patent-leather shoes; her hair was artfully piled on her head and fastened together under a black velvet hair bow. But as beautiful as she looked that night, the thing that would be etched forever into my memory was the look on her face. I'd never seen Ellie look quite that way. She was without question the prettiest little girl in the world—more beautiful than I'd ever realized. It was her radiance that overwhelmed me. Her face shone with poise, confidence, excitement, delight, and anticipation. Like Ellie, I got so caught up in the fairy tale that I temporarily lost all my bearings. At that moment, I didn't even stop to consider where the clothes had come from. And my earlier disapproval was as remote as summertime.

I wanted nothing more than to hug her and wish her well. But for an instant, I felt intimidated. Yes, *intimidated* by my own daughter. I wanted her to come my way so I could take a closer look. I wanted her to rush up and call me Daddy and ask me what I thought.

Sure enough, she and Emma did head my way, and I felt my heartbeat surge. As I stood there, watching them approach, I thought back to when Ellie had left for the first time to spend the night with Emma. The girls were four then. I remembered how she clung to me, how she said she didn't want to go anywhere if I couldn't go, too. I told her then that I'd always be with her—no matter where she went. She told me how she could hardly wait to get back to tell me all about it. Back then, my approval meant the world to her. But that was long ago.

This time, she walked regally across the street and down the sidewalk toward me, a self-possessed girl, no longer little. It was not hard to imagine the worldy, sophisticated woman she would someday become. At her side, Emma looked dazzling in a burgundy velvet dress with a white lacy collar and cuffs.

I considered what to say. I wanted it to be something witty, something memorable, something to let her know I wasn't mad anymore. Something to make light of the whole situation. I would tease her with a line like: "Found something to wear at the bottom of your suitcase, did you?"

But Ellie and Emma breezed right by me, without so much as a hello, whispering in that conspiratorial, nose-in-the-air way that girls and women do around men who hold no interest for them. They stepped into the camper to say good-bye to Patti. Out there on the sidewalk with my trees, the pavement felt cold under my feet, and my arms felt as limp as sawed-off limbs.

The next thing I saw was Ellie back on the street corner, throwing her arm up into the air, like a veteran New Yorker. A cab was curbside in an instant. She, Emma, and Anne scooted in and, just as quickly as it came, the cab vanished into a line of traffic.

"Ellie," I called out, wanting more than anything to say good-bye, but a honk from an impatient cabbie behind them drowned me out. As I stood there on the corner, it occurred to me that Patti was wrong about one thing: Ellie had not left under the dark cloud of my disapproval. She was off to have a big night on the town—the biggest night of her life.

5

The Long Night

I pulled out my trusty pocket watch and checked the time. I'd wait up for her. Surely she'd be back by nine-thirty or ten at latest. I smiled to myself, thinking about how the sequence of events would play out. Ellie would come back from her adventure dying to tell me what had happened. She'd be so full of the experience that it would override any other emotion, namely, having felt hurt or slighted by me. She was still enough of a kid to want to tell all *immediately*. I've noticed in life that once someone tells you his or her story and you really listen, your differences fade away. It's just like getting stone-faced customers to laugh.

I resolved then and there to cast aside all my reservations and enter into the spirit of the occasion. Usually when Ellie told me about an adventure, her excitement would get me going, too. She'd start telling me what happened, and I'd interrupt with questions. Tonight, I'd ask: Where did you sit? What did you think of the show? Of Lincoln Center? And the outfit? Was it a loan? Did you go out and spend all your money on it?

She'd generally answer my questions, then jump around from place to place, getting confused and backtracking to explain this and that. Finally, I'd suggest she start over and tell me what happened right from the beginning. Then I'd impose a rule on myself—no questions or interruptions until she had finished.

This time, it would just be the two of us. With Patti and the boys in bed, I'd have Ellie's undivided attention, and she'd have mine. Maybe if she were still going strong, I'd suggest we go to the all-night coffee shop around the corner and make it into an occasion. We'd order dessert or herbal tea and just be together. I might even say something self-deprecating about my behavior that would make her realize that I was sorry for not having signed on to the spirit of her adventure earlier on, that I was concerned I'd been too harsh.

If I checked my pocket watch once that night, I checked it a thousand times. I pulled it out every five minutes, watching the slow creep of the dial. I did this even more frequently after ten o'clock, when our tree business died. It was cold and windy, and every so often I'd step into the camper and make myself some hot tea to warm up. But when the family turned in, I stayed outside longer than usual, sweeping needles and pacing up and down the sidewalks to keep moving.

It was approaching midnight, and still there was no sign of Ellie. The cordless phone, which I'd slipped into my jacket pocket, hadn't rung all evening. What if something had happened to her? My mind jumped to worst-case scenarios: Might she have been kidnapped? Since the Abbotts lived just one block above us on Eighth Avenue, which ran one-way uptown, the cabbie might have let them out on their corner. Could Ellie have told Anne Abbott that she'd just walk the block from Emma's to our camper by herself? Could Anne have allowed that?

Might something dreadful have happened to her en route or elsewhere? Ellie's confident tone was so impressive that adults often forgot how young she still was. As I flashed on images of my daughter in trouble—a victim of a random shooting, in the clutches of some pervert, or hurt or wounded along the side of the road—sweat poured from my body. Not long before, I'd read about a child who was standing on a street corner and had been hit and killed by a city bus that ran up on the curve. I felt powerless, knowing that if she were in trouble, there was nothing I could do to help.

Then I remembered: Ellie had gone into the camper to speak to Patti before she left. Probably she'd asked permission to spend the night with Emma, and Patti had failed to mention it to me. One of the ironies of this season is that Patti and I are both so busy, sometimes days will pass before we have any private time with each other. It was logical that since things had been so tense between us, Ellie would have turned to Patti instead of me.

So I did something unprecedented. I delayed counting the day's sales total; I'd do it the next day. I stripped down to my long thermal underwear and climbed into bed. I reached over and touched my wife's long brown hair, spread out like a fan above her head on the pillow. Sleeping soundly, Patti rolled over and opened one eye.

"Did Ellie ask to spend the night with Emma?" I asked. She didn't hear me the first time, so I had to repeat myself.

"She didn't say anything about it to me," Patti said. Then both her eyes opened as she seemed to come fully awake: "Isn't she here now? What time is it?"

I answered her questions and speculated that she must have gone to Emma's.

Patti thought for a moment. "I'm sure you're right. Ellie's got to be at Emma's. Maybe they got in so late that they didn't want to disturb us—or maybe they aren't back yet." Patti turned back on her side and mumbled something about hoping Ellie had had a great time. With that, she shut her eyes and slipped back to sleep.

I had brought with me into our bed the cordless phone, which I laid alongside the pillow. I held it to my ear and pushed the "on" button, just to be sure it was working.

I slept fitfully that night. I kept waking to check my pocket watch and my daughter's bunk, as if she might have magically reappeared. But all I saw was Zippy, looking forlorn with his monkey arm dangling off the side of

Ellie's bed. By the wee hours, it should have been obvious that she wasn't coming home that night, but that didn't stop me from looking. More than once, I considered calling the Abbott home despite the lateness of the hour, but my better judgment prevailed.

I woke up for good just after five that morning—early for me. Since Patti and the boys were trying to sleep, I made an effort to dress quietly. But the camper is so small that it's hard for anyone to do anything without everyone sharing in the experience. I wanted hot tea, but the kettle would start whistling louder than the most persistent alarm clock. Instead, I twisted the cap off a bottle of water and sat on the bench seat below Ellie's bunk, thinking. Zippy's arm, dangling down from above, touched my head.

As I sipped, I could hear New York City waking up. Though the early-morning traffic hadn't yet started its rumble, the garbage men had already come and gone, and the delivery trucks were going strong, unloading the many goods needed to keep the city running. Diesel-fueled trucks were throwing off bundles of *New York Times* and *Daily News* newspapers, which dropped with loud thuds onto the sidewalks in front of three delis within hearing distance. Delivery men off-loading newspapers, soft drinks, bread, and canned goods paid no attention to the hour of day, and their halting exchanges were shouted in decibels exceeding their noisy engines. With our camper right on the street, the sounds of the city were amplified for us.

As I tallied up the totals from the previous day, I could feel my left eye pulsing its nervous twitch. The totals were good, but the twitch was a sign that I was getting tight and tense and needed to get out and stretch. I stepped outside into the pitch-black world around me, slipping the phone into my jacket pocket.

As I walked, I couldn't stop thinking about Ellie. Perhaps I'd been wrong in assuming that if we lived our lives with integrity and made friends with as many of our neighbors as possible, we would live in a kind of safe bubble on Jane Street. I've always said that you don't build security through locks and barriers but rather through treating everyone decently, with dignity and respect. As I found myself strolling in the direction of Emma's apartment building, the voices of my Vermont friends asking me if I was worried about bringing my family to the city came back to me. At the time I brushed them off, but if something had happened to Ellie, I'd have only myself to blame for being so casual—and cavalier—about her safety.

I looked up at the apartment, scrutinizing Emma's bedroom window for some sort of an answer. I punched the "on" button on my portable phone and was about to dial the number I knew by heart. Realizing that it was not yet half past five, I restrained myself. I would have to wait. But the dark windows didn't offer me a clue.

I had returned to the stand and stayed busy plumping the branches on the trees. I was keeping my eye on the time. At seven sharp, I dialed their number. The phone was answered after two rings by Anne, who sounded groggy.

"I was calling about Ellie," I started. "Is she there?" I didn't want to let her know how worried I was, so I consciously tried to tamp down my anxiety.

"Oh, sure," she answered breezily. But, she told me, Ellie was still asleep, "recovering" from her big night on the town. Something in Anne's tone carried a whiff of disapproval, as if she thought I had been riding Ellie too hard. She didn't offer to wake Ellie, but said she'd send her "along home" after breakfast.

I clicked off the phone, overcome by relief. I wanted to dance a jig. Ellie was safe! Ellie would be back!

I returned to my trees. Though I initially felt grateful and relieved that Ellie was very much alive, something began bothering me. I recognized it as irritation. No, the feeling was stronger than that; it was *anger*. I had spent one whole night agonizing when one simple phone call from her would have put my mind to rest. It wasn't like Ellie not to call. The Ellie I knew was conscientious and considerate. For her to spend the night out and not call, or go out and not think of calling, didn't fit her character. Was she changing so rapidly that I didn't know her anymore?

This much was clear: Ellie and I needed to have a serious talk.

I had just finished picking up a scattering of trash that had blown onto the stand from points unknown during the windy night before—you know, things like advertising circulars, empty Styrofoam cups and boxes, and plastic grocery bags—when I spotted her. I watched Ellie step down from the stoop at Emma's building. Although it was a fairly short distance from Emma's to the stand, she was taking her time, peering into shop windows and even bending down to read a newspaper in a box. Dressed in her denim jumper and work boots, she looked as plain today as she had elegant the night before—Cinderella after the ball.

When she finally arrived at the stand at half past nine, I had known for more than two hours that she was safe and sound. But I was steaming. Even though I already knew, and she knew that I knew, I had to ask. "Where have you been?"

She rolled her eyes. "You know. I spent the night at Emma's. Anne told me you called this morning. At *seven.*" The way she emphasized the word made it clear that she viewed it as an unreasonable hour.

I looked at my daughter. "And who gave you permission to spend the night out?"

"Well, no one, but I figured you'd know where I was."

"How could I know where you were? I was worried sick about you all night."

She didn't say anything, but one of those looks crossed her face like she wasn't really there. So, to get my point across, I continued. "I am amazed you would think it was okay to just not come home at night without calling to let your mother and me know where you were. And since when do you spend the night at someone else's place without first asking permission?"

"Well, I could hardly ask your permission because you weren't even speaking to me yesterday."

"I wasn't speaking to *you?*" I heard myself say. It was happening again—I heard myself shouting. "It seems to me that you weren't speaking to *me*. I don't know where or how it's come into your mind, but it appears that you think you're running the show right now."

"Daddy, you are a—" she started, as if she were going to come up with some insulting epithet but thought the better of it.

"So you had it all worked out—*before* you went to *The Nutcracker*—that you were going to spend the night at Emma's. Do I have this right?"

"No, I hadn't decided *before*. It's just that we got in late. The show was late and dinner was late and it just seemed like the thing to do. I didn't really think it through."

The way she said—"it just seemed like the thing to do"—got to me. It sounded so casual, like someone else talking. It sounded like her responsibility to us was insignificant, a bother almost. I had no other choice: I was simply going to have to exercise my parental power.

This was one of those moments when a father needed to come on strong.

"Ellie," I said in a stern tone that set the stage for what was to come. "You've spent more time at Emma's this season than you have here. But you've spent all the time that you're going to over there. I need a little help at the stand." Although it might sound harsh, I wasn't about to reward her after she had been so irresponsible. Ellie needed to learn a lesson. "You're not going to go to Anne Abbott's party tonight."

Her eyes widened in disbelief, and then the red flush of anger spread across her face. "Daddy," she said, "I promised them I was going. I *promised* them I would help them get ready."

"You *promised* me you were going to help and you haven't much this year."

Her shoulders drooped and the spark in her eyes went flat. Looking defeated, she just stared at me, saying nothing.

"Look, Ellie," I said. "This stand is not just for fun; this is our *livelihood*." But she wouldn't understand, and the sullen look hardened on her face. "Not only are you not going to the party. You're not going *anywhere* from this stand without my permission."

Then she raised her hand, like some sarcastic school kid. Her voice was strangely pitched when she asked: "May I go into the camper?"

I nodded my head yes, and she marched into the camper and closed the door behind her.

I got busy with customers. Henry was with me and even sold a few candles for his absentee sister. When I had a break in business and ducked my head inside the camper, I was expecting to see her at the table, reading or writing or making last-minute Christmas cards. This time, I was prepared for her pouty behavior and one-word answers. But to my surprise, Ellie wasn't there. Neither were Patti and Timmy.

When Patti and Timmy returned a while later, I asked Patti where Ellie was.

"I thought she was with you," came her reply. Patti looked distracted, and I could sense she was impatient with my ongoing conflict with Ellie. "Maybe she's at Emma's."

"She couldn't be," I said. "I grounded her."

"Grounded her—Billy, what is going on? You've never grounded Ellie in your life. I've never even heard you use that word—*grounded*. This doesn't happen in our family. Billy, *what* is going on?" She unclipped her barrette and gathered her hair back into one hand, rearranging it nervously. It was easy to see that she wasn't upset with Ellie but had somehow transferred all the blame to me.

I had the feeling then that I'd had as a kid when my sisters ganged up on me. I knew I was in the right, and that I had no choice but to hold my ground. But for the rest of the day, I was haunted by a strange sense of loss.

6

A Stranger in the Family

When I returned from walking Santos early that evening, I saw that a man had taken my place at the stand. He was handing a tree to a customer and pocketing the cash. I didn't recognize him, so my heart started beating fast.

But when I got close, I realized it was Mark, one of our Jane Street friends of long standing, a gold trader on Wall Street. Every year, I heard the same refrain from him. "You've got it made, Billy. Great family! Three great kids! Beautiful wife! I'd trade places with you in a heartbeat."

And he did, every year—for a night.

Mark was walking up and down my sidewalks, proprietarily fluffing up the branches of the open trees, rear-ranging the wreaths decked out along the fence, humming "Jingle Bells." When he saw me, he challenged: "What are *you* doing here?"

"I could ask you the same question."

"Go get ready," he commanded, handing me the keys to his apartment. "Go take a shower."

"Ready for what?" I asked.

"The party," he said. When he could see I was clueless, he explained. "Patti asked me to fill in tonight for the party. What—she didn't tell you?"

"Oh, Anne Abbott's Christmas party," I said. "I forgot."

A wicker picnic basket overflowing with candy canes, Christmas cookies, and fruitcake sat on Ellie's abandoned candle stand. Two shiny thermoses bracketed the basket like bookends. "Something to keep me going," he explained. "Something to keep my customers happy." He winked when he used the word *my*. "Like a cup?" It was clear that for Mark, retailing was a kind of pleasure sport.

I declined. He poured himself a cup of hot coffee, took a sip, and slapped me on the back. "You know, Billy, you've got it made. Every year, it just gets better for you. What you have lasts a lifetime."

Mark couldn't have known all I'd been through these past few weeks with Ellie, but he told me exactly what I needed to hear. He looked at my life the way I liked to see it myself—the way I hoped it would become once again. "I appreciate your offering to help out tonight," I said, "but I'm really not up for going to this party. So take it easy tonight, or if you want to sell trees, work with me."

"I don't care if you're in the mood or not," Mark responded. "You're going to that party. What are you— crazy? Patti works like a dog all month and all she asks of you is to go to one party. You go. You don't ask, you don't consider how you feel, you just go."

It may have been Mark's glowing perception of my life that sent me off to shower, shave, and spiff up for the party like an obedient, young boy. Or maybe it was his bluster. In my heart, though, I was doing it for Patti. Mark just told me what I already knew and needed to hear. She *had* been great this year, as usual. Working long hours without complaining. Cooking in a kitchen that even a saint would have to describe as modest. Remaining steady and sensible no matter what storms raged around her. Patti never raised her voice. Don't get me wrong. Patti was no pushover, and once she staked out her position, she held her ground. So when she asked for something in a certain tone—when she said, "humor me"—I took notice. Patti used that tone on me when she saw me talking to Mark and came rushing out of the camper.

"April has taken the boys to the party with Heidi and Anya," Patti said, referring to Mark's wife and kids. "You and I are leaving in twenty minutes."

If my heart's not in a thing, I've always had a hard time just going along with the program. Even though I'd agreed to go for Patti's sake and had spiffed up as she requested, my feet were dragging. Patti and I were walking along Eighth Avenue to the party when we got into it.

I was taking my time, staring through a plate-glass window into an antique store all decked out in Christmas finery. Bolts of red and green velvet fabric draped Victorian chairs, tables, and rockers, with vintage Christmas cards propped all around. A merry family of porcelain dolls dressed in satin coats and fur muffs sat by a simulated fireplace, as if they'd just come in from the cold. The entire scene was meant to convey a feeling of gaiety, holiday cheer, and to carry you back in time.

But rather than bolstering my spirits, the tableau had the opposite effect on me. What I really felt was empty and hollow, like I'd been left behind by the season, left behind by my own family. Sure, Christmas tree sales were meeting—even exceeding—expectations, but it was hard for me to think of anything but Ellie and how sad I felt. She seemed more interested in a world that I knew nothing about—a world in which I didn't live—than in our life together as a family. I wondered if, in hindsight, we should have come to New York at all. I hadn't broached the subject with Patti but I was considering whether to make this our last season on Jane Street.

Patti tapped my shoulder. "Billy, pick up your feet. We're already late. Let's get going."

I looked at her, knowing I had to level with her. Even if she didn't accept my reasoning, at least she had to know what was on my mind. "I don't really want to go to this," I said, wrinkling my forehead. "You know that. I'm dog tired. It's been a long, hard season and I haven't had much help." I felt railroaded, the way that I did when I was a kid and had to offer grace at the dinner table because I was the only boy among five children. When Dad called on me, as he did for Christmas dinner, it always made me feel self-conscious. I felt the same way in these clothes—a red shirt, wool pants, and suspenders—my own version of fancy. There was little I hated more in life than being forced into a role I didn't feel, forced to make pleasant conversation and be gracious when I was down. All I really wanted was to get back to the stand and sell more trees. Clean out the stock. My Christmas wish was for things to return to normal in our family, and I didn't see how going to this party was going to accomplish that.

Patti stopped walking and pulled on my arm to stop me as well. It was her signal that she needed my undivided attention. "Billy, it's December twenty-second. There are two full days before Christmas, and I want them to be peaceful, happy days." She didn't point any fingers but the implication was clear: she viewed me as an obstacle to family harmony.

"The old 'peace-at-any-price' message," I heard myself saying. "That's what you're trying to sell." I know it sounded sarcastic, but it just came out.

"I'm not trying to *sell* anything," she responded. "I'm just asking you to grow up. Be a man, be a father."

"Me—grow up?"

"You. Ellie's got to go her own way, forge her own path. And not you or anyone should try to stop her."

"This is not about forging a path in life," I said. "This is about her breaking rules and not having consideration for us, her parents. I grounded Ellie, and she ignored me. It may sound old-fashioned, but she *disobeyed* me." Just thinking about how she'd behaved fueled my anger. "What I ought to do is go over to the Abbotts, collect Ellie, and bring her home."

"Billy, you're doing nothing of the sort." There was that tone again.

"But once you start bending the rules with children, once you let them run the show," I protested, "you're lost. Patti, we've been over this. You can't let your kids openly disobey you. You've always agreed with me on this point."

"I do agree with you—in principle—but this time I think *you're* the one who's being unreasonable. You don't ground someone who's worked as hard as Ellie has all season for not calling home one night. It was late. She didn't want to wake anyone. She's a kid, Billy. She got swept up in the moment. She never expected us to say no. You've got to show your child that you trust her. That's the most important thing."

"She has to show that she's worthy of trust," I replied. "She hasn't done a very good job of that. She wasn't thinking about the fact that I stayed up half the night worrying about her. We're a family. Rule number one in a family is you've got to show others consideration."

"She's a *kid*, Billy. She got caught up in the moment."

"Well if that's the case and if she gets away with it, that sets a dangerous precedent."

Patti gave me one of those searching looks as if she was really trying to find an answer in my face. "Billy, how could you even *think* of forbidding her from going to the big Christmas party? It's just about her favorite part about Christmas. She looks forward to this party all year."

As we walked up the stairwell to their apartment, silence parted the two of us. I didn't understand why Patti couldn't see my position, didn't see things the way I did; evidently, she felt the same way about me. But from my experience, I knew that sometimes you have to draw a hard line. "She's already been to *The Nutcracker*, which preoccupied her all month. Now she's wrapped up in going to Christmas parties. My life is not about supporting Ellie in some fairy-tale fantasy, Patti." I cleared my throat. "Life is not a fairy tale. This is not how we can afford to live."

Patti's breathing had grown markedly heavier—not from the stairs we were climbing but from something she was wrestling with inside. When we arrived at the Abbotts' landing, she took a deep breath, as if to bolster her confidence. Then she said: "This is all about fear, isn't it, Billy? You're afraid of losing her."

"What are you talking about?"

We were standing right in front of Anne Abbott's substantial door, decorated with a Balsam wreath, one of ours. Fancy chocolates, wrapped in bright foil, dangled like ornaments from the wreath's boughs. Conversation from the party mingled with music and came out in waves through the cracks around the door.

Patti's jaw was set. "I'm not walking in there with you acting like this. You're thinking of yourself and only yourself." When she said this, I recognized that at some level it might be true. But her comment only served to make me feel even more estranged from the party, from Christmas, from myself. Standing there just then, I felt like a stranger in my own family.

When I made no response, Patti continued. "You can't do this to your daughter. You're tearing her up, Billy."

My voice was small: "I don't think I'm having any impact at all."

"Ellie is capable of putting on a good front. Just like the rest of us."

"I don't think we're seeing any front. She's made herself clear. She wants to be Ellie Abbott."

"Billy," she said in a way that caught my attention. For the first time that night, I really looked at my wife. Patti is a slim, petite woman who always looks youthful and energetic. But tonight something was different about her—her eyes were dashed by a streak of sadness. "Billy," she repeated, sounding exhausted and defeated. "Just go home. With this attitude, you're going to spoil the party, and I don't even want you to be here."

Her words stung. As I stood there, stunned, an image of my father from many Christmases ago flashed before me. When he was diagnosed with tuberculosis, my father was sent off to live year-round in a sanitarium, where he was supposed to regain his health. He was allowed to leave for special occasions, like Christmas. But when he came home for the holidays, when I was about Ellie's age,

I remember that he seemed more like a visitor to our home than the head of the household. As I thought of him, it struck me for the first time that even though we didn't live apart, I, too, was becoming a stranger in my family. Ellie and Patti were going around me, making their own plans, plans that didn't involve me. Perhaps Patti was right—at the heart of my conflict with Ellie was my fear of losing her.

Before I was able to respond, the apartment door swung open. Anne Abbott was at the door, showing out guests. "Merry Christmas to you!" she said, pecking them on the cheeks. "Give my best to your folks!"

Behind her was their formal oak dining table, covered with a fancy lace tablecloth. A giant spread—enough food to feed a proverbial army—lay on the table, including an enormous turkey with all the trimmings, ham, cranberries, baked apples, vegetable dishes, and breads of every variety. Monogrammed napkins and bone-china plates were neatly arranged on one end. Guests traveled around the table nibbling on the food, heaping their plates high, and talking loudly to one another. An amazing array of desserts—cakes, tarts, cookies, nuts, and sweetmeats—sat on a sideboard.

"Patti, you look wonderful!" said Anne, hugging my wife. "Billy, welcome," she said, a bit more tersely, before motioning us inside.

Beyond the dining room was a spacious, high-ceilinged living room. Holding center stage was a huge, magnificent Douglas fir tree. A mound of wrapped presents sprawled out around and beneath the tree, isolating and protecting it. Standing before two long windows, the tree flaunted its beauty to the street.

An enormous crystal punch bowl, surrounded by delicate cups, occupied its own table in the living room. When I looked over I saw Ellie, dressed once again in the same magical outfit as the night before. She was merrily bantering with guests of all ages—an eclectic Greenwich Village mix—and playing hostess by ladling up eggnog. One of the guests, an older man who looked and sounded Italian, said to Emma, "Where have you been hiding your delightful friend?"

"She lives in Vermont," Emma explained. "She's just visiting."

This time, I recognized Ellie instantly. Though she looked as dazzling in her black velvet jumper and lacy white blouse as she had the night before, having seen the entire effect once, I wasn't as startled by it. Still, as I stared at her, I realized that I wasn't fully accustomed to this Ellie either—Ellie, the poised young lady, no longer a child.

"Billy Romp," called out a tree customer who quizzed me about the season. "Had a good one?"

I told him that I had.

"How is it stacking up compared to last year?" he wanted to know.

"I can't complain," I said, not wanting to get into it.

He reported that the tree he'd bought from me was the prettiest ever. "Did they get their tree from you?" he asked, nodding at the Douglas fir. When I glanced back at the tree, I saw that Ellie had vanished. After my customer and I parted, I began moving around the elegant apartment, trying not to engage with anyone, looking for my daughter.

I'd been in the apartment before, of course. I'd been to Anne's Christmas parties in years past. But I'd never set foot in Emma's bedroom. It has always struck me as off-limits, inappropriate somehow for a man to be entering the territory of a little girl who is not his daughter. Ellie had described Emma's bedroom as being off a circular stairway. So when I came to a circular stairway in the hall-way behind the kitchen, I decided to give it a try. Quietly, I climbed the stairs, not wanting to startle anyone. I crossed my fingers that I'd find Ellie up there, so that the two of us could have a private talk.

The landing at the top of the circular stairway opened to just one door. That door was ajar, so I could peer inside without stepping in. What I saw of Emma's room took my breath away. An elegant cherry canopy bed with frilly white eyelet bedding and matching curtains looked like a boudoir fit for a princess. Floor-to-ceiling shelves overflowed with dolls like I'd never seen before. For an instant, I wished that I could afford something like this for Ellie. It occurred to me that girls' attraction to this fantasy world was similar to boys' fascination with sports and hunting. I flashed on my own boyhood fantasies about the wilderness experience.

I heard girls' voices: Ellie's, Emma's, and several others I didn't recognize were coming from inside the room.

"*The Nutcracker* really *was* amazing," I heard Ellie say-ing. "Of course, I loved the dance of the sugarplum fairies. And the Nutcracker when he came to life, and the costumes and the ballerinas and all that." The tone of her voice took a turn when she asked the others a question: "But do you know who my favorite character was?"

"No. Who, Ellie? Who?" one of the girls said.

"Dr. Drosselmeyer, the godfather. Because he's a clever inventor who can make anything. He made an eye patch from plaster and a wig from spun glass. He reminds me of my dad. Dad carries his toolbox with him everywhere, and he can fix anything—everything that no one else can, that everyone else has given up on."

"Do you have to fix more things when you live in the country?" another girl chimed in.

"Of course, you do, silly," Emma answered for Ellie. "No one has a superintendent up there. Everyone has to do for themselves."

Ellie told the girls about all these projects that I worked on in my spare time back home—the woodshed that I'd put up, the fences I repaired, the furniture I made from salvage pieces from condemned houses. She talked about how I'd bought the very camper that we live in when we come to New York for peanuts. "Dad said it was held together with chewing gum and bailing twine. He worked on it for one whole year—putting in new lights, a roof, and redoing the plumbing—making it livable again."

"You live in a camper!" one of the girls exclaimed. "You mean, right here in the city?"

"Yeah—right down the street!" Ellie answered.

"It's right on the corner of Jane Street and Eighth Avenue," Emma added. "You know, where all the trees are."

Ellie proceeded to describe life in the camper. How she slept on a bunk that was really a large shelf over the dining room table. How every day she and Henry lugged water in gallon jugs from Bonsignour across the street. She talked about her mom's homemade biscuits and how Santos was around to protect us, but he never had to because everyone on Jane Street was so kind and loving. As I stood there listening, I was entranced by the life she was depicting, the life she was leading. My heart was turning somersaults as I realized, standing there, that Ellie was not running away from our life together. It was clear that she loved our simple life and would bring it with her into every new world she entered.

"Tell them about the candles," Emma said. I knew I had no right to continue eavesdropping like this, but I couldn't seem to tear myself away. I felt like a dry tree that's just been put in a pan of water. I was absorbing Ellie's words like water, and they were feeding my soul.

"Well, this year I went into business for myself," Ellie said, sounding like a veteran entrepreneur. "I've learned a lot of stuff about business from my dad so I decided to sell candles. And guess what—in less than three weeks I made $327.50!"

I was touched, overwhelmed, and delighted by what I was hearing. Ellie had spunk, drive, and enough self-confidence to pursue her dreams. Even though I hadn't been able to give the material things she might yearn for, I realized that I had given her something as valuable—the tools to make her own way in life. I saw in an instant that despite my fears, my daughter didn't want to be Ellie Abbott of Manhattan after all; she was happy as Ellie Romp of Shoreham, Vermont!

I retraced my steps down the stairs and tried to find a place to do some serious thinking. I saw a side room, went inside, and shut the door. I stepped over to a wall mirror and ran my fingers through my hair. It was true that the season had been demanding and that we were all tired, but that was no excuse. I had to own up to it: I had seriously misread my daughter. Even though I saw her constantly, I had fallen dangerously out of touch with Ellie, just as my father had with me and my sisters when we were children. If it continued this way, it would become a strain for my own children to talk to me, just as it had been for me as a boy to speak with my father. I thought of what Patti and Ellie had said about me and realized how poorly I must have behaved for them to come out like that. *I* had been the one with the problem that year.

Time was running out on the holiday season—and the year for that matter—but, as Patti had mentioned, there were two full days before Christmas. There was still time to make amends. And already, ideas were churning.

A soft knock interrupted my thoughts. As I stepped toward the door, it inched open. Next thing I knew, I saw Ellie peering out from behind it. At first, she appeared startled to see me. "Dad," she said sheepishly. "I didn't know you were here."

"Wouldn't have missed this party for the world." I felt cheerful and slightly mischievous. Ellie looked at me, searchingly; I could tell she recognized that my mood had changed—and maybe more than my mood. I led her to the mirror and stood her in front of it, she in her elegant outfit and me in my suspenders and wool pants. We weren't dressed alike, but I'd never felt closer to Ellie than at that moment. I laid my hands on her shoulders, feeling the soft velvet of her jumper under my hand. "Ellie, you look wonderful," I said. "Just like a princess."

She beamed, not saying anything, not wanting the moment to end.

I touched her velvet hair bow, which set off her hair so beautifully, then moved my hand to a resting spot on her back. It occurred to me then that she might want to stay on at Emma's, maybe until we left for Vermont. I saw then that it was good for her and, for the first time, I actually wanted her to stay. "You'd probably like to spend the night with Emma. And you have my blessing. You always have my blessing—even when I'm acting like an old grouch."

"Oh, Daddy," she said, stretching her arms around me. "Thank you. But I don't *want* to spend the night here. I want to come home." She looked impish. "I miss Santos."

"And what about your candles? You've got to keep an eye on Henry, or he'll move in on your territory," I teased.

Together, Ellie and I walked into the living room to

find Patti. Arm in arm, we were the picture of happiness. When Patti saw us, she could tell instantly that I'd come around, as I generally do. "Are you ready to go?" I asked. "Or do you want to stay awhile longer?" Whatever she wished, I would have gladly obliged.

Patti sunk her eyes into mine and then she smiled. "Let's head back."

Patti, Ellie, the boys and I took the long route back to the camper. Our pace was slow and leisurely; we were in no hurry. We drank in the sights and smells of Christmas all around us. We savored every minute together, a family once again.

When we approached a corner florist, I reached into a bucket and plucked out the prettiest long-stemmed red rose. I paid the clerk and told him to keep the change. I dropped to one knee and offered it to my wife. "Patti, for you," I said. "For my beautiful wife. For putting up with me."

She raised the rose up into the air theatrically. "Here's to a wonderful Christmas! It starts this minute."

Once again, Patti was prophetic. It turned out to be the beginning of the most wonderful Christmas of my life.

7

Our Conspiracy of Kindness

I don't know where the idea came from but I suspect it had something to do with my reconciliation with Ellie. While loading a beautiful ten-foot Balsam into the vehicle belonging to a couple who lived in the neighborhood but spent every Christmas at their country house, I had an impulse not to charge them.

"On the house," I told Donna and Mark. From the way their eyes glistened, it was obvious they were touched. Suddenly I became possessed by the idea of taking it one step further—shutting the stand early and distributing my wreaths around Jane Street, not for money but for love.

Trees and wreaths had been flying off the racks all morning. On Christmas Eve, customers rarely lingered to chat. They made up their minds in a snap, pointing to a tree and putting cash on the barrel head in one continuous motion. Actually, Christmas Eve can be surprisingly busy. Any other season, I would have tried to squeeze out last-minute sales, like the racer who can see he's won but keeps pushing to improve his time at the finish line.

But this year was different. This year, I was overcome by a celebratory feeling, like I'd come to a crossroads and made a turn that was profound and significant. I know it might not seem important to someone outside the family, but I see now that my struggle with Ellie over her going to *The Nutcracker* was really a struggle with myself. It was a struggle over what kind of parent I was going to be—a gatekeeper or a guide, a disciplinarian or a friend. On this Christmas Eve, I felt relief, like I'd made the right choice, having almost chosen wrong. It was similar to how I felt when I first fell in love with Patti—I was grateful, gleeful, and shaken up inside, almost like I'd been reborn.

This sense of connectedness with my family extended to everyone I encountered, people I spoke with and those I saw, like the pedestrians rushing up and down Eighth Avenue, their large shopping bags overflowing with tubes of wrapping paper, gifts, and Christmas greenery. I cast this warm, protective beam of love on my daughter, who was busy helping customers. Gone was the separation between Ellie and me. We were back together again, working as a team, for the family good.

The camper was filling with gifts delivered by the good folks of Jane Street who'd left them along with their farewells. Because space was so limited, Patti would graciously accept their offerings, place the presents in shopping bags, then make regular runs to Laura's apartment across the street to store them till departure. That cleared the floor for visitors and family and kept the presents safely out of the boys' reach. Timmy and Henry's excitement was already so great that they could barely contain themselves.

Because people knew that our day was so hectic—and that we were leaving—Patti never had to cook on Christmas Eve. We were always showered with food, including a complete Christmas dinner "to go" from Bonsignour, wrapped in tin foil and delivered to our camper door by Philippe himself. Mike and Horton from the Jane Street Tavern provided sandwiches for the drive. Angela dropped off homemade cookies. Renato, the owner of Piccolo Angelo restaurant, came bearing chicken, pasta, Italian bread, and cheesecake. Our little camper overflowed with edibles—enough to feed all our visitors.

When I stepped in, Richard and Debbie, a caring couple who lived adjacent to the community garden, were sitting at the table talking with Patti. They'd brought a picnic basket from Balducci's overflowing with bread, cheese, fruit, juice, pastries, and muffins.

"I can't believe you're leaving us," said Debbie, who was handing Timmy cheese in pieces the size of almonds. "You're a fixture here on Jane Street."

"We love being here," I said.

"When you're here, somehow, I feel like all's right with the world."

What Debbie said was touching. It brought home a belief I've long held—that each person has a tremendous impact not only on those with whom he or she comes into contact, but on everyone with whom *those* people come into contact. Ultimately, even the smallest of actions are far-reaching; ripples travel the world. I didn't know how to respond.

She continued: "We can count on your coming back next year." There was a note of doubt in her voice, which made me wonder how she happened to ask that question on this of all years. Ever since the party, my notion to stop coming to Jane Street had vanished; in its place was a longing to return, even before we had left.

"As far as I know, yes, we'll be back."

"With bells on," Patti chimed in.

"Bless you," Debbie said. Before leaving, she pecked Patti and me on both cheeks, European-style. After she and Richard had stepped outside the camper, she poked her head back in and took one more wistful look around our modest abode.

Buoyed by Patti's blessing to give away any wreath and every tree I wanted, I found Ellie outside. "How would you feel about going around on a mission of great importance?" I asked.

Ellie looked puzzled. "You mean, leaving the stand?"

"Pick out the prettiest of the wreaths, add some extra ribbon and pinecones, and let's hit the trail."

It took her maybe three seconds to decide to accompany me. "Did you get an order?"

I smiled mysteriously, transferring a hammer and nails from my toolbox into my jacket pocket. "You could call it that, but not from a paying customer," I said.

She looked puzzled but intrigued. The old Ellie once again, she tugged on her braid, considering the larger picture. "But who's going to mind the stand?"

I found a big cardboard sign, a red Magic Marker, and wrote out, "CHRISTMAS TREES—FREE TO GOOD HOMES!"

She read the sign then looked at me, incredulous. "Dad?"

Just as I was scanning the stand for a prominent position in which to place the sign, along came Louie, our homeless friend. Louie usually arrived at the camper between ten and twelve to see if there were any leftover biscuits or muffins from our breakfast. At that moment it dawned on me—here was an opportunity for him. "Louie, would you like to sell trees?" I asked.

"Anything to help you out," he said. He removed his cap and scratched his balding head. "How much would I make?"

"Whatever you charge," I said.

He didn't get my gist. "Just tell me what you'll pay," he said, "*after* you take your cut."

I repeated myself: "Whatever you charge belongs to you. Every dime."

"Good deal," he said, but still looked puzzled. But when a customer approached, Louie went right to work. Maybe he thought he'd work out the details later.

"Dad, is it okay if I bring some candles along?" Ellie asked. I could tell she was entering into the spirit of our scheme, our conspiracy of kindness. When I agreed, she loaded her candles into a heavy-duty canvas carrying bag.

Henry saw that we were up to something and charged out of the camper.

"Henry," Ellie jumped in, "if you sell my candles, you'll make whatever you charge!"

"How much do I have to give you?"

She repeated the magic words: "Whatever you charge belongs to you."

We started on Jane Street, west of Eighth Avenue, my favorite section in the West Village. A series of lovely old brownstone town houses flanked cobblestone streets. The brownstones had large stoops and small yards defined by wrought-iron fences. Some of the yards boasted in-ground trees, majestic and delicate, old and young. Christmas lights had been strung onto some, including the now-leafless deciduous varieties.

When we found a brownstone that lacked a wreath—there weren't many—we climbed the deep granite steps and sprang into action. Working quickly to avoid detection, Ellie would hold up the wreath as I tacked it to the door. If there was a brass knocker, we'd position it in the center of the wreath. On the top step, to the side of the doormat, Ellie would place one of her candles. We left no note, no message, not even our familiar hand-lettered "Romp Family Christmas Trees" card that we used to tag trees that had been sold or were waiting to be delivered. On one brownstone we passed, the wreath had fallen and lay on the mat. Without saying a word to each other, we climbed the stairs and restored it to its rightful place on the door.

As Ellie and I walked along the neighborhood, carrying wreaths and candles, people smiled and waved, mistakenly assuming that we were delivering wreaths that had been ordered. Full of the Christmas spirit, some stopped to talk.

We bumped into the happy new father who'd bought a Christmas tree from us the night that Ellie had gone to *The Nutcracker*. He was carrying a turkey and other fixings for a big holiday meal. Holding his hand, his step-daughter was carrying a poinsettia, decorated with green foil that pointed up around the rim of the pot like leaves. Usually I can remember a face and have trouble attaching a name, but that little girl had made an impression and her name stuck—Erica, the angel. Seeing the wreaths dangling from my arm, her stepfather asked to buy one.

"Not for sale," I told him. He looked disappointed, then quickly recovered his good cheer. "That's all right; I'll pick one up at the florist around the corner." He was surprised when I handed him a wreath.

"I thought you said they were all spoken for," he said.

"They are, but not for sale. This is our gift to you and Erica, on your first Christmas together." He gave me a look as if I were amazing to have remembered something so significant to him.

As Ellie and I traipsed around the neighborhood, we both got more and more consumed by the Christmas spirit. We became giddy with giving. Our heads became light, and we seemed to float down the street. Looking back on it now, I have to admit I'd never had more fun with my daughter before or since. Somehow on that very day I'd shed the heavy mantle of merchant, the responsibility of having to come out ahead in every exchange, and had replaced it with the pleasure of giving for giving's sake. I made a discovery: Christmas was not about getting and spending or, in my case, about exceeding my previous year's sales figures. This is what the cynics would have you believe, what I myself had thought in the past. But it exists in your spirit, your mind, and your heart. *You* make it happen by giving to those you encounter—those you love and those who are hard to love. You make it happen by loving unconditionally.

During our "stealth mission" I gazed into the windows of the first-floor apartments, trying not to be too obvious. Many of the tastefully decorated Christmas trees I recognized as my former charges, trees that I'd brought to Jane Street and sold off the stand. Under most trees, brightly wrapped gifts had already started to accumulate. In one home, a woman clad in an evergreen-colored cardigan sweater with gold buttons shaped like trees was stringing red yarn through cookies. Once they were firmly anchored on sugar-cookie wreaths and strapped along the chests of gingerbread men, she'd dangle these treasures from the branches of her Fraser fir. In another window, I saw a man nailing stockings to an intricately carved mahogany mantelpiece, while his wife hung pine boughs all around. Three young children, two girls and a boy, played a board game on the floor.

Ellie followed my gaze into the worlds behind the windows, and I sensed the longing in her eyes. During our mission together, I'd become more attuned to my daughter's thoughts and desires. In the past, I had always assumed that what I wanted would satisfy her, as well. But when I glanced at her that time and found her transfixed by these homey scenes, I began to wonder. Could she have tired of the commerce of the season and be wishing we could just be a family, like other families, celebrating Christmas Eve together? She had said as much earlier, the first time she brought up the subject of *The Nutcracker*. Could it be that her attraction to the ballet was based on the mystery and magic of the story line itself—something that was lacking in our practical, workaday lives?

For the past ten years, we'd always left the stand around midnight on Christmas Eve after a full and busy day. The drive to Vermont took five to seven hours, depending on the weather; we generally drove through the night, arriving at our home before dawn. But just because that was the way it always had been didn't mean that was the way it always had to be. I hatched an idea—something different, a kind of extension of Ellie's and my Jane Street "romp." But this particular "conspiracy" was aimed at my own family's happiness, especially Ellie's. This idea, coupled with the gift for her that I'd been working on furiously, I hoped would make this Christmas stand out in her mind the rest of her life.

We would leave early this year. We would spend a special Christmas at home in Vermont. If we left in an hour or two—that very afternoon—we could enjoy Christmas Eve at home. Maybe some hot cider and carols in front of a blazing fire. Then we could all sleep through the night in our very own beds, wake up on Christmas morning, and take baths or showers in our very own tub, and not drag ourselves into the living room sleep-deprived and weary from a long night's drive. The only loose end was to finish Ellie's Christmas present. The moment of its unveiling was fast approaching, just hours away. I had my work cut out for me.

When we'd given out the last wreath and candle, I said: "Let's go home, Ellie."

"To get more wreaths?"

"No," I said. "Home—home to Vermont."

When we got back to the stand, Louie was gone and Henry was dickering with a man who wanted to buy the corner tree, the one that Henry had picked out so excitedly just four weeks earlier. Henry was telling his customer that he'd have to charge extra because it had ornaments on it. He scratched his head and named a figure. "Maybe thirty-five dollars."

I took Henry aside and explained the new order of things. He went back to the customer and told him that we'd closed up shop. "We're not selling anything anymore, just giving stuff away."

"You mean for free?" the man asked, incredulously, as if such a thing were almost un-American.

"Well, he didn't say anything against tips," Henry added.

Even though it was a bit peculiar, given that it wasn't even five o'clock, the kids settled into the backseat of the truck wearing their pajamas, just like always. They loaded in pillows, blankets, and snacks for the long ride home. Timmy tied a red bandana around Santos's neck. Henry was still counting his money. Clutching Zippy, Ellie listened raptly for the first sign that our wheels were rolling so she could repeat what she said every year: "So—we're off!"

Even in the Village, traffic was heavy as we left Jane Street heading toward the West Side Highway. We'd take that road to get out of New York City, then point our truck toward the Hudson River Valley. We'd keep going all the way to Lake George, and make the short jump over to the Champlain Valley in Vermont and then home to Shoreham. Driving for the first time in weeks, I took it slow and easy in Manhattan, careful to avoid potholes and aggressive drivers. After all, the truck was carrying precious cargo: my family and our Douglas fir tree, the one we would set up at home, the one that would be our honored guest for Christmas.

After having disappeared for an hour or two, Louie was there at the stand when we left. Through the rearview mirror, I watched him talking animatedly with a new customer. The last thing he'd told me, after attempting to give me some money for the trees, was that he'd given a price break to some "poor people" who'd come by the stand. Then he smiled a big smile through a mouth short of teeth. "You take care of that family," he said, tears welling in his eyes. "You got a good one. The best."

I thanked him for reminding me and told him I would try extra hard and I'd keep trying for the rest of my life. It was a solemn and heartfelt promise, one I vowed then that I would never break.

8

Christmas in Vermont

Patti and I woke up before dawn on Christmas morning in our very own bed in Vermont. The first thing I did—before brushing my teeth or changing out of my pajamas—was to rush down the stairs and throw open the front door. It had been thirty years since I started my Christmas this way, but when I was a kid, I observed this ritual zealously. Every Christmas morning—before checking out the bulges in my stocking or surveying the presents under the tree—I would dash outside to see if it had snowed and whether I could spot any reindeer or runner tracks left by Santa's sleigh. I don't know what possessed me that morning to be a kid again, but that's how it happened.

I should say *tried* to throw open the front door. The fact is it would not budge. You see, I was pressing against a mound of snow. As we'd pulled into our driveway the night before, light flakes had begun to dot the windshield, but only then—the following morning—did I realize that the flakes had amounted to a storm. I shoved the door open and looked outside. "Patti," I called. "Come look."

Patti padded over to me in her house slippers, carrying two steaming cups of tea, a specialty blend from Jane Street friends. We stood at the door, sipping our tea, staring at what lay before us. Snow had draped the evergreens around our house and formed a virgin carpet on the lawn. In the distance, only the Green Mountains and the gray sky of dawn interrupted the great white swath covering miles of fields and farms and homes.

Snow blanketed every barn, fence, and house in sight and acted as a great equalizer. It camouflaged all flaws and unified the landscape into one harmonious whole. It brought people together, too, just like my Christmas trees did in New York City. Unplowed, the road in front of our house was piled with two feet of snow, with higher drifts at the fences. Save for a few power and telephone lines, there were almost no visible signs of civilization in this rural Vermont countryside. You'd half expect to see a horse-drawn sleigh filled with carolers happen along, dragging a tree behind.

Though white Christmases were not unusual in Vermont, a Christmas Eve snowstorm was. "Can you believe this?" Patti said, looking at me incredulously. It was more of a comment than a question. She and I were both thinking the same thing—what if? What if we hadn't left New York City when we did? What if the storm had come earlier, when we were en route? It was a blessing that we'd made it home when we did. If the timing had been different, we might still have been on Jane Street, or, more likely yet, stuck at some welcome center, somewhere, marking time with other stranded travelers.

"I'm so glad we're home safe and sound. Tomorrow, after the excitement of Christmas wears down, we can shake out the needles from the season and get on with it," I said, pulling her close. "Shaking out the needles from the season" was our private family expression for this brief period, the aftermath of Jane Street. It was a cozy, lazy time when our family flopped together on the couch, caught up on our reading and mail, talked and giggled a lot, and felt happy and relieved that the season was finally over. It was also a time of readjustment to the country and of letting go of the excitement and energy on Jane Street.

As the children slept, Patti and I prepared our home for the big day ahead. Patti set the few remaining candles from Ellie's stock on the dining room table and placed one on every windowsill downstairs. I shoveled a narrow path to the camper and, when I stepped inside, saw that our most precious cargo had weathered the journey safely. Before leaving Manhattan, Ellie and Henry had selected a Christmas tree—a lovely, bushy Douglas fir— and wrapped it in a favorite old quilt that had once belonged to their grandmother Gilmartin. They wanted the tree to rest comfortably on the ride home. When I lifted the tree from its protective covering, I saw that it had shed few needles and that all its branches remained intact. I hoisted the trunk, Patti took the midsection, and we carried the tree into the house. On the stoop, I sawed off a thick slice of the tree trunk, knowing it would be the last cut of the season.

Back inside, working as quietly as possible so as not to disturb the children, Patti and I set up our tree in the stand. We hung ornaments, many of them handmade through the years by the children out of dry macaroni and cardboard, pinecones, cotton balls, and construction paper. We took special care with older pieces from Patti's mother, hanging them on high boughs, out of the reach of Timmy, Santos, and Patches.

When Patti was in the other room, I removed from my toolbox a special package I'd brought from the city. I speared open the cellophane overwrap, lifted out its contents, and began hanging its long, silvery strands on the branches. When Patti returned with several shopping bags full of presents, she stopped short in amazement.

"But you don't like tinsel," Patti said. She was talking slowly, at half speed, as if she were making some kind of mental adjustment as she spoke.

"*You* do," I replied.

I made my way out to the shop to put some finishing touches on my gift to Ellie. It was finished, really, except for some last-minute buffing. As I worked, I couldn't help but wonder if Ellie would like it and understand its message.

When I was satisfied, I placed my gift for Ellie, cushioned with tissue paper, into a sturdy box—the old-fashioned kind made with two distinct parts, a short top and a deep bottom. I went through a stack of newspaper cartoons looking for the right strips to put on top. The Romp family tradition of wrapping gifts in "the funnies" began as an economy measure in the sixties, but several of my sisters and I kept it going long after we had homes of our own and could afford real wrapping paper. It was our touchstone with tradition, our way of celebrating our past while linking it with our future. To complete Ellie's package, I tied the package with some beautiful evergreen-colored grosgrain ribbon that I'd found at a specialty shop near Jane Street.

I placed my gift on the floor, in an obscure spot in the corner behind the tree. I was hoping that Ellie wouldn't notice my present until all the other gifts had been unwrapped, and all the excitement had died down. This way, the present would command her full attention.

The kids slept until the sun shone full force into their windows. But once they opened their eyes and remembered that it was Christmas and that they were home in Vermont, it took them seconds to dress and come bounding down the stairs. Ellie dashed up to me and gave me a giant hug. "Merry Christmas!" she said, her eyes sparkling in anticipation of what lay ahead. Just as quickly as she arrived, she ran back upstairs, returning with another shirt on—a red flannel, almost identical to the one I was wearing. To correspond with my blue jeans, she had put on her beloved denim jumper.

Fewer spectacles are more enjoyable than watching your children open presents on Christmas morning. I tell Patti and the kids every year that I would be happy receiving no gifts at all because my pleasure lies in watching *them*. (Of course, every year they ignore my plea.) The children started with their stockings, which were stuffed with oranges, walnuts, candy, small trinkets, and at least one tool for each child. There was a hammer in Henry's stocking with a note attached: "Santa says that next year you'll be old enough to nail the lights to the rack on Jane Street."

For the first time that Christmas, little Timmy caught on to how our gift-opening ritual worked. Each family member took his or her turn peeling off paper, with the rest of us acting as audience. Timmy would patiently wait his turn, but when it came, he held nothing back. He'd rip the paper covering his present to shreds and wad it into a launchable missile. Once he emptied a box, he'd stomp on it and shriek if he could make it pop.

More seasoned in the ways of Christmas, Henry extended his role at center stage. When his moment came to unwrap, he would prolong the tension by pulling off the ribbon and removing the paper *slowly*. Then, when he discovered the item in a box or bag, he'd react with animated facial expressions, but delay producing it, so everyone was left to wonder for a moment what it was and whether or not he liked the gift. Always entertaining, his theatrics were worth the wait.

Ellie opened her gifts thoughtfully. She used scissors to cut the tape where the edges of the paper overlapped, then after removing the wrapping neatly folded the sheets and slid them into a bag for reuse the following year; she detached the ribbons and bows from the package and put them into yet another bag. She finished all her housekeeping before getting to the reward.

I could tell by the way her eyebrows arched when she first saw my gift that she knew it was from me. Perhaps reading my mind once again—or perhaps planning it all on her own—she saved my gift for last.

When the moment came, Ellie tugged at the ribbon and raised the box top. She peered inside, still not knowing what to expect. Then she lifted out my present, handling it as if she were a museum curator unpacking some priceless, irreplaceable treasure.

When the tissue fell away from what she was holding and the present was revealed, her mouth fell open and her eyes met mine. Was it disappointment on her face, or some other emotion? It was hard to tell. Clearly, this was not what she had been expecting. At that moment my daughter reminded me of how Patti had reacted earlier that morning to my putting tinsel on the tree. Like her mother, Ellie seemed to be taking in not only her present but a new version of me.

"Oh, Daddy," she said at last. "A nutcracker!"

The toy nutcracker I'd carved stood twelve inches tall. He was garbed in complete dress uniform: a fancy red helmet and matching jacket with gold epaulets. Several impressive military medals adorned his jacket, and there were gold bands at his cuffs. He had black hair, a curly mustache, a goatee, and a nose that would have seemed exceedingly prominent if it weren't for his mouth. Oversized, his mouth was unified by a deck of sharp pearly whites that carried just a hint of menace. Lifting the nutcracker's coattails activated his jaw. I'd tested him out a number of times: stick a walnut in his mouth and pull his coattails, and the shell would crush, liberating its meaty center.

Examining the nutcracker as thoroughly as an obstetrician checks a newborn, Ellie worked the nut-cracking lever up and down. "How did you ever?" she started, gazing up at me in wonder. She had enough experience with wood carving herself to know how hard my job had been, and just how difficult it must have been to keep my project a secret on Jane Street.

"Oh, it was nothing."

"And I thought you were making me something else. Like a toolbox."

"That was *my* dream for you, not yours," I said. "Once I realized that, I scrapped my plan and decided to make what you wanted."

It is said that change never happens overnight. And that may be right. Perhaps genuine change works its way through your system gradually—chipping away at barriers in your unconscious and making breakthroughs when you least expect them. But it sure didn't feel that way at the time. It felt like I'd been struck by some powerful force that night at the party, transformed into a new person. Then and there, I vowed to part company with my old self. I decided to give Henry the gift he was coveting, the one I'd been making all year for Ellie—the toolbox. I decided to make Ellie the thing she wanted most, a nutcracker. The gift was meant as my blessing for Ellie to pursue whatever dreams she dreamed, even if they departed from my own. As long as she had it, the gift would remind her of her big night at *The Nutcracker*, and all the obstacles she'd overcome to get there. With that gift, I'd made a promise to myself to stand behind her wherever she might go in life.

Ellie was sitting on the couch in front of the tree when I squatted before her, my face about level with the nutcracker's. I was one of those people who had a hard time apologizing. I would know in my heart when the need was there, but it was never easy to express in words. I took a deep breath and then the plunge. "If I made it hard on you this year, I want to tell you how sorry I am," I started. "I've always known that I should let you be your own person and pursue your dreams. But, in my heart, I felt that if you went places I wasn't going, you were leaving me behind."

Tears welled in Ellie's eyes as I spoke. "When you didn't want me to go to *The Nutcracker,* I thought you were mad at *me.*"

I reached for my daughter and held her tightly. "Ellie," I said. "No matter where you go in life—no matter what you do—I'll always love you and support you. I consider you a miracle, always have, just as I did the day you were born. This nutcracker is here to remind you of how I feel. I'm only human, so even if I mess up, he'll carry the message; he'll never forget."

Just as suddenly as her tears had come, they were gone. "Know what?" she said brightly. "Next year when we go to Jane Street—if I sell twice as many candles, maybe three times as many as this year—and I can afford it—" She paused, studying me intently. "Could I, I mean, would you go if I paid to take you and the family to *The Nutcracker?*"

I didn't answer her immediately, but a smile broke out on my face, so she could tell I would offer no objection. Ellie wasted no time spinning out a new dream.

"Even Timmy—he'll be old enough then. I'm just dying for you to see it. Especially the tree."

"You don't think I see enough trees in what I do?" I

kidded. "You're going to take me to Lincoln Center to see a *tree*?"

"But this tree is like no other," Ellie started. She could tell I was teasing, so she kept on, becoming more extravagant as she proceeded. "It grows like Jack and the Beanstalk. It almost comes alive onstage, kind of like our trees."

"I wouldn't go just to see a tree, but I *would* go to see the ballerinas and the dance of the sugarplum fairies. And I'd definitely go to see you all dressed up in your fancy black velvet outfit. If it still fits."

"Well, if I'm buying the tickets, you'd better be dressed right."

"You mean, you wouldn't take me as I am," I said, feigning indignation and snapping one of my suspenders.

"No way," she said. "You'd better start getting your outfit ready. Coat, tie, dress pants, the works."

I haven't worn a tie since my wedding. Or was it my high school prom? In the past I would have raised an objection—told her I wouldn't go because I didn't feel comfortable dressing up. Or it might have been a matter of principle, my problem with spending so much money on some fleeting pleasure. I would have found a reason.

But at that moment, I knew better. I would put on a coat and tie for my daughter. I would take an evening off from the stand. I would be by her side, cheering on her dreams and decisions, pitching in to help wherever life took her.

Sure, I was the man of the house and took pride in my role as provider. And just as I'd hoped and prayed at the start of the season, tree sales *had* broken all records that year, getting us over a major hump financially. But I knew that Christmas morning that I'd remember this season not for what I'd made but what I'd learned.

Ellie's gift to me that year was showing me how to be a good father. Not that she set out to teach me a lesson, or even knew in advance what I needed to know. It just worked out that way. I'd always thought I understood what being a father was all about. My version had included spending lots of time with my children and teaching them the fundamentals. And those are good things. I hate to admit it, though, but I see now that I'd gotten a little smug somewhere along the way. Just look at how great my children were, I said to myself! How bright, how considerate, how well behaved and well liked they were! In hindsight, I realize that, in Ellie, I'd seen a miniature version of myself. It had been easy to love her that way. It had been harder when she was behaving like someone else entirely—"someone" I wasn't comfortable, or familiar, with.

My struggle with Ellie made me think about my customers. They came back to my stand year after year—and considered themselves my friends—because they liked the way I treated them. I gave them rope, I let them be themselves. It's strange how you can have a special gift like mine and apply it so effectively in one area of your life but so ineffectively in another. I had turned on my talent for people with my customers but not with my family.

The gift I received from Ellie that Christmas was a lesson in how to give without expecting *anything* in return. Not a tree sale, or a helper, not someone to take care of me in my old age. Once I really "got" the gift, once I really understood it, I was overcome by a feeling of headiness. It felt a bit like giving away the wreaths and trees on Jane Street. I learned that Christmas that what I needed most was to learn how to give for the sake of giving. Once I'd learned to give Ellie away—to give her to herself—I had embraced the true meaning of Christmas.

For some reason deep within her, Ellie needed to go to *The Nutcracker* that Christmas. And ever since she'd gone, I'd watched her come alive like my Christmas trees after I snip the string. Ellie had told me that first day on Jane Street that you learn by trying new things. And she was right. She needed to learn by testing her wings. That was precisely what she was doing. And now she had my blessing.

Ellie was holding the nutcracker in her arms when suddenly she jumped off the couch and took center stage on the floor. From the way she was standing, holding her shoulders back and her back erect like a ballerina, I wondered if she weren't imagining herself in the part of Clara onstage at Lincoln Center. She thrust the nutcracker up in the air and then did a perfect pirouette on the toe of her tennis shoe. As I watched her, she appeared graceful and ladylike, looking once again more like the woman she would become than the tomboy she still was.

Catching her fire, I jumped up to join her where she was standing. "I'll go to *The Nutcracker* next year!" I said, wanting with my whole heart and soul to go. Suddenly the opportunity to see *The Nutcracker*—this ballet that had changed my daughter's life—seemed like an enormous privilege, one of life's greatest. "Our whole family will. We'll all be there together!"

Ellie gazed up at me, her eyes radiating happiness, anticipation, and the ultimate contentment of a child who's well loved.

"Once we shake out the needles from this season," I said, "we'll have an entire year to get ready for the big night!"